Shadows on the Hill

D.G. JARDINE

THE REMARKABLES STATION QUEENSTOWN

A.H. & A.W. REED

Wellington Sydney London

First published 1978

A. H. & A. W. REED LTD
65-67 Taranaki Street, Wellington
53 Myoora Road, Terrey Hills, Sydney 2084
11 Southampton Row, London WC1B 5HA
also
16 Beresford Street, Auckland
Cnr Mowbray and Thackeray Streets,
Waltham, Christchurch 2

ISBN 0 589 01009 3

Typeset by Typocrafters, Auckland
Jacket design by Fraser Gardyne
Typography and layout by Kevin Woolcott

Printed by Kyodo Shing Loong
Printing Industries Pte. Ltd

This book is dedicated to my wife —
and a beardie collie dog.

I see again the upland wilds,
Stern, rugged, bleak, and bare;
The strong winds sweep o'er the hill sides steep
And the tussocks toss in the icy air
Silver and gold in the changing light,
Gold and silver far up on the heights
Of the mountain wild and bare.

David McKee Wright

CONTENTS

PART ONE

Tales of station ways and station days on Kawarau Falls

PART TWO

PREFACE

THESE mountains that cradle the blue waters of Lake Wakatipu are admired by all who visit the area — but what visitors do not at first appreciate is that these mountains, in common with almost all those on the eastern side of the Southern Alps, are the base upon which rests New Zealand's unique pastoral industry of high-country sheep farming.

When this is explained the first question asked is, "However do you get the sheep down?" — These few sketches endeavour to answer that question and convey something of the methods used in managing a high-country station.

The author apologises for the use of non-metric references, but he has as yet not adjusted to recognising the metric equivalent of, say, a horse of fourteen hands or a couple of feet of snow.

Acknowledgments are gratefully made to Mr P. M. Chandler, Ettrick, Central Otago, for much historical data, and to the directors and staffs of the Hocken Library, Dunedin, and the Lands & Survey Department, Dunedin, for their help.

Also, the following books have been consulted: C. J. Griffiths, *W. G. Reeves*; W. A. Taylor, *Lore and History of the South Island Maori*; Elsdon Best, *Polynesian Voyagers*; David Lewis, *Wind, Wave, Star and Bird*; F. G. Hall-Jones, *Historical Southland*; John Hall-Jones, *Mr Surveyor Thompson*; K. C. MacDonald, *History of North Otago*; Alfred H. Duncan, *The Wakatipians*; Brett's Historical Series, *Early New Zealand.*

Remarkables Station
Lake Wakatipu

D.G.J.

THE REMARKABLES — "There are many larger and higher ranges, but few more unusual or strikingly beautiful."

1

THE PACK TRACK

By valley, shoulder,
Bluff and hill
Old packtracks
Beckon onward still.
They mark the path some lives have led
And point the way to what is said.

THERE must be many of us who, bored and tired from a long car journey on a hot summer day, have pulled to the side of the road in some high saddle of our tussock-clad inland hills, seeking a brief respite from that hypnotic, shimmering, black bitumen ribbon.

Switching off the engine and flinging open the door we step from the confines of a hot and stuffy car and, all unknowingly, break through the cocoon of artificial man-made environment with which we have come to surround ourselves and to accept as the norm, into the true environment of the land.

Our first impressions are of the flow of clean cold air pouring steadily over the saddle and the silence of nature in repose that lies over the apparently limitless extent of country under our eyes. Below, the distant brown plains are fretted with the green of farms, with a bright reflected sparkle here and there to mark lake or stream. Beyond the saddle, below a sky of deepest blue, the tawny-gold hills recede in ever-rising ranks to underline the snowclad peaks, barely discernible in the blue haze of distance — for we are on the doorstep of the mountain lands.

As a bonus to all this we are aware of one of the great joys of being an island nation — that there, not so very far beyond our horizon, the rolling waters of the Pacific or the Tasman are bursting on rocky headland or caressing the sands of some protected bay. But here it is the mountains that hold our attention, gentle saddle and clearcut peak, beckoning us on to find what the misty valleys hold.

As we return reluctantly to the car, how many of us note the faint trail that crosses the saddle beside the road? It is worthy of our notice and a little thought, for there lies the way, the pack track of the pioneers.

It is perhaps fair comment on our modern life that we should pass in our thousands over our motorways and leave no trace, yet here, almost lost to sight among the tussocks, there yet remains mute evidence of the trials and fortitude of the pioneers — an insignificant yet appropriate memorial to those first few who laid the foundation of a new life in a new land.

These tracks, which once played such an important part in back-country life, are being superseded by roads. As time passes, more and more become unused, overgrown and indistinct — they belong to an era which even now is fading in the rising mists of time.

In places they may yet be clearly seen — deeply cut in the jutting shoulder of some bold spur or winding along the bank of a chuckling stream. Tracks worn deep over the years by the passage of man and horse, of thousand upon thousands of sheep, aye, and cattle too. They were the arteries along which flowed the burgeoning life-stream of human progress in a new land; stockmen, packers, shearers, women and children, goldminers, swaggers, all who would pass this way have, of necessity, trodden or been carried by the indispensable horse over the same track — each, in passing, helping create, this their monument, in these silent lands.

How varied must have been the loads that were transported over these winding trails — food, timber, pipes and machinery, cherished pieces of furniture, wool, even children in panniers on some trusted packhorse. (The packers seldom failed to devise a way in which the most unlikely loads could be carried.)

The people whose passing formed these tracks would have more than their share of journeys filled with hope and despair, joy and sorrow too, for many passed this way but once — storms and illness took their toll and the unmarked graves are not a few, or simply "Somebody's Darling". If it were possible, what tales might be sifted from the dust of these old tracks. . . .

The roads of today, naturally, become familiar to those who use them often, but there is not the same personal association to be enjoyed in their use as was the case with the old pack track. Though the main tracks were maintained by the Counties, few indeed of those who used them would fail to pause to remove some obstructing rock that had rolled down, or divert an errant stream that was doing damage.

Those who have known the back country will certainly, in memory, often follow these winding tracks, either swiftly on the back of some well-remembered, free-walking horse, or slowly at the tail of a long string of sheep. He who used them will recall days when he drove himself on, bent against the force and cold of a blizzard; or summer days, when the heat beat strongly down into the gullies and the crickets sang in the tussock along the way — the welcome shade of the beech groves and the clean, cold taste of the creeks that crossed his path. The pauses to "boil up" for the

Here the pack track is the beach of Lake Wakatipu. The date is 1931, the packer D. Garforth.

ever-welcome billy tea; and, not least, the companions who shared the way.

Very much a part of life in the old station days — as many of us know who have followed them from one block of high country to another — they led to new country where fresh experiences often became cherished memories. They hold a rightful place in any tale of station days; in many cases they provided the only connection between adjacent areas, neighbouring stations, or "the flats". Now, their days of usefulness drawing to a close, they remain a visible reminder and a link with past experiences and associations.

As a pack track has been, for many, the way that was first followed into the back country, so it has been used here to lead into this collection of sketches, snatched at random from the turning pages of memory "along the way".

Some will be passing through old familiar country. To them is extended an invitation to share once more, in imagination, the satisfaction of a day on "the hill". To those who have had no active association with the mountains, may these few sketches be of help in appreciating the life of a high-country station.

2

FIRST DAY ON THE HILL

The Spirits that
In mountains lie,
Are intertwined twixt
Earth and Sky.
As bent the twig shall grow the tree.
Who knows the hills shall ne'er be free.

HE was tall for sixteen years, almost gangly, the faded blue dungaree trousers just that much too short above the fairly new hobnailed boots. The old grey woollen school shirt was obviously becoming too tight here and there. With the two dogs beside him he leaned on his hillstick and watched Steve, the head shepherd, and Murray, the other musterer, top the back of the basin and disappear from sight. He felt a surge of satisfaction and accomplishment — at last he was doing a man's job, he was a musterer, albeit of only four hours' experience.

From where he stood among the upper limits of the snow tussock, Brown's Basin rolled away to the long sharp ridge that formed the main divide. Here and there a tarn glittered under the just risen sun. To the north, in sharp contrast, towered the south face of the Double Cone, which shot up like the point of some broad stone dagger from the earth, thrusting aside the fallen boulders at its base; doomed to receive only minutes of sun in the year, the sheer face of raw rock was almost black in this early light. From small crannies here and there and from its foot, amidst the scattered mass of black stone, came the pale blue-white gleam of old snow.

At this altitude, of over 7,000 feet, the clear morning light left no soft edges to rocks or shadow. All was sharply etched, almost unreal in its clarity — golden tussock against dark shadow, black rock against pale green sky. He sensed, for the first time, that here was a different world far removed in time from the sunlit plains and valleys below. Here one was conscious of the power and forces of creation that still seemed to flow, undiminished, unsoftened by the passage of time. His gaze passed out into the void of space beyond the peaks, hundreds of miles of misty valley and plain, only now receiving the first edge of the sun. Behind, as far as he could see, stretched the snow-covered Southern Alps, dominated here and there by the major peaks.

The south face of the Double Cone, ". . . doomed to receive only minutes of sun in the year . . .", height 7,680 feet.

With almost conscious effort he transferred his attention to the opposite side of the valley, the Lumber Box, which they were to muster that day. Still mostly in shadow it looked a mass of bluffs and he tried to visualise the beats the others would take and the track Steve had said the sheep would follow.

He tried to recall what Steve had said last night when he had given them each their beats for the day — Doug, Cliff, and Jim on the Front Face, themselves on the Back. Murray top on the Back, Steve through the middle (where, I don't know, it's all bluffs), you on the bottom. Doug top on the front (be sure to make contact with Murray at Rough Creek saddle), Cliff next (if he can't find Goat Pass go underneath), Jim on the bottom and come off together or they would lose the wethers back through certain bluffs, like last time when Doug went to sleep (indignant protests of innocence from Doug and a summary of the ancestry of certain persons who haven't got enough wind left to yell out loud enough to be heard). More details, questions and answers that he did not fully understand.

However, he recalled clearly what Steve had said before he and Murray left him: "You wait here while we go over the back. We'll hunt the basin back out to you, so keep quiet. When you get behind them push them on a bit then go on down to the foot of the Bridal Veil'' (the long lacy falls below) ''and wait for us to come out from below the Cones. Keep a bit behind me all the way down to the valley.'' Well, that was easy enough, so he climbed to the top of the short spur to see across the basin.

Two lines of sheep were coming rapidly towards him across the basin, wending in between the tarns. As they reached the middle of the basin they formed into one long line heading straight for him. He dropped quietly down the hill into the high tussock and sat waiting concealed with his dogs for them to come round. After about five minutes there came a rattle in the shingle above and a big wether, closely followed by others, belted across the spur not ten yards away.

Oblivious to all else, the sheep were intent on putting as much distance, in as short a time as possible, between themselves and the dogs which had started them on their way. They had soon passed and needed no hunting to keep them moving. Two small mobs on the face above dropped down and joined the string, which was soon well down the valley, with others small grazing mobs moving quietly off in the same direction.

Away out across the basin, along the foot of the big ridge from the Cones, he spotted the two tiny figures of the others. With a string of sheep in front and a short string of dogs behind, they moved steadily across the broken face, heading for the opposite side of the valley. There was no sound of dogs and he remembered that Steve had said he and Murray would not really have the high mob till they got between them and the Queen's Drive, a pass running across high up on the Front Face.

Turning away he dropped down through the tussock beside the Bridal Veil to the floor of the valley, to wait for the others to come round.

The sun now had some heat in it and he had a drink from the creek. It was ice cold, crystal clear, but singularly tasteless — bred of rocks and snow it had had no time to absorb the flavours of the lower streams. He sank back into an ''old man'' snow tussock, unslinging the tuckerbag from his back. Steve had shown him last night how to make it; take an old sugarbag, tie the ends of a bit of rope or your spare bootlaces to the bottom corners, a half-hitch over the neck and there you are. The only trouble with them, Steve said, was that your sweat can sometimes get right through into your lunch and spoil it. The lunch, in its broken newspaper wrapping, looked battered but good, two of bread and meat, one of jam and a slice of cold duff. He ate half of it, closely watched by Old Jed and Roy who received a crust each.

As he watched them gulp down their share he recalled what his Dad (''The Skipper'') had told him when he had given them to

"The Skipper". In 1922 Dickson Jardine, the author's father, took up Kawarau Falls Station, of which the Remarkables Station is a part.

him: "They've forgotten more about mustering than you know at present; work them quietly."

Roy was a big, smooth-coated, black-and-tan huntaway who, although old, was spoken of with respect by the other musterers. He had been broken in by a head shepherd who was now managing a place. Old Jed was a black curly-haired heading bitch, bred by the legendary Miller, with a mind very much of her own. This she had displayed that morning on the way up, when Steve had told him to pull a small mob down to the flat to "see what your heading dog is like"; she pulled too hard and nearly put them over a bluff. Much to his discomfort, Steve had told him to be sure and feed her that night, as he didn't much care for dogs that displayed an interest in creating their own dogtucker.

As he settled back into the big snow tussock to wait, the events of the past two days passed through his mind — the seven-mile walk out to the hut with the others — the bustle of settling in, attending to packhorses, killing mutton and dogtucker (he recalled with satisfaction that he had held his own at that job, thanks to the number of times Steve had given it to him to do this past year). The candle and firelit evening in the hut with Darby the packer sorting stores and preparing for the morning, the others either in their bunks reading, or hammering new nails in their boots. The hard but not quite uncomfortable straw mattress — the pungent smell of pinched-out candles and the slow flicker of the fire — then a roar from Darby had wakened them to a quick breakfast of chops and toast and steaming black tea before the crackling fire. Cut your lunch and out into the dark (it was 2 am), stumbling through the bracken to let the dogs off.

A fight had broken out amongst the thirty-odd dogs milling about, but had been quickly quelled by the swipes of several hill-sticks. The dogs were fit and fresh now, looking for a scrap: after a couple of days on the hill there would be none of this, just the odd snarls from those that had formed a mutual dislike.

The other three started off for the Front Face and Steve led off straight up the hill. Though it was pitch dark he seemed to have no difficulty in finding his way through the bracken and scrub and set a pace that did not vary whatever the pitch.

After a steep forty-minute climb Steve had angled across the face. From the deeper blackness below came the distant rumble of the creek and all about them was the blackness of the hills, faintly outlined against the glittering stars. As they topped a rise they were met by an icy drift of air and he had realised they were entering the mouth of the Wye Valley. In dark folds it stretched before them, but to his surprise for there was no moon, bathed now in a faint light. He had been puzzled for a time till he had realised the light came from a bright star, away out over the head of the valley. He came to know it later as Venus and it was to light his way on many another moonless morning.

It was just breaking grey behind the peaks when they crossed the creek. A short steep pinch and the whole valley floor lay ahead, hemmed in on either side by soaring bluffs and spurs all bathed in the grim grey of dawn, but far up at the valley head soared the peaks of the Double Cone, pink-edged with the first rays of the sun.

The creek rumbled quietly below to the whispering overtone of a gentle breeze in the snowgrass, the sun beat down warmly and he slept, for all of five minutes. With a start he awoke and, in horror of having committed the most heinous of shepherding crimes, leapt to his feet, startling the two old dogs who had been snoozing beside him. How long had he slept — had the others gone past? Hell, he'd let them down and made a blue on his first day. Against the rumble of the creek he strained his ears for the sound of dogs or irate call, while his eyes searched the face above — nothing moved. After five restless minutes he heard faint but clear and miles down the valley the sound of dogs — he was away with a rush and was nearly a mile downstream when he saw sheep grazing high up on the face where the others should have gone through. He

"His gaze passed out into the void of space . . . ." The author, a musterer in the making, 1928.

stopped dead and tried to figure it out, listening hard for any indication of where the others were. He knew the rule that if you sang out and barked up, the man on the next beat was supposed to answer, and he had just filled his lungs to shout when, from high up in the bluffs above where he had left, came a clear long call and a barking of dogs that echoed through the valley. What relief! As he started back up the valley he wondered about the sound of dogs further down which he was certain — almost — he had heard. He was to learn later that in the living silence of the hills any sound sought long enough can be heard, from "way out there". He was not the first to be fooled and would not be the last.

Steve had watched the whole performance and knew what had happened. He grinned over the hurrying figure on the valley floor and swung in behind the string of sheep now drawing out through the bluffs.

They moved steadily through the face, top man in front, middle man slightly behind and the boy on the bottom slightly behind him, now paying strict attention, making plenty of noise and for the first time sensing the combined effort and underlying purpose of the day.

High up in the bluffs the string of sheep increased in number, hurrying along their well-known track, along this ledge, over that bluff, down that particular chute and by early afternoon were out on to the relatively clear final spur. He now had a small string in front of him. Some had been hunted down, some had been pulled down by Old Jed to join the string. She knew her job all right, anticipating a break this way or their desire to run for that blind ledge in the bluff, not paying any more attention than necessary to the over-numerous commands of her young master.

There came a halt of ten minutes or so while the top man waited in Rough Creek saddle for the top man on the Front Face to come round, then they moved on. The sheep were mobbing on the point of the spur ahead, caught between the two gangs; they moved restlessly from point to point, on down to the last chute, down which they now came pouring like a wave, raising a huge cloud of dust as the men from the front and back closed in. A thousand or more big fit wethers went surging through the gate into the holding paddock.

They headed down to the hut, to the waiting meal of stew and rice and dried apricots, tea and bread and more tea. Over the meal the talk centred round the events of the day — the run that such and such a dog had made, the small mob that got bluffed, the keas were working on the Front Face, the big treble fleecer that had been caught away from his favourite bluff and ended up in the paddock, and back to the run that some other dog had made.

They turned into their bunks now, it was 3 pm, to doze till suppertime at 7 o'clock, feed the dogs and back to the bunks.

The boy stretched out in complete satisfaction. His legs were a bit stiff and his feet a bit swollen (take them down to the creek he was told, but later). It was good to lie down.

3

EARLY DAYS

Child alone
In a world of men,
Ever seeks their
Ways to ken.
Lessons learned are learned full well.
Experience needs not school or bell.

THE 1920s saw the decline and, to all intents, the end of the era for the big goldmining schemes.

The public of the time had not the varied and attractive avenues of "investment" that are so ably presented today — no TAB, finance companies, motel/hotel syndicates, building societies, or even the SEARCH FOR OIL. The goldmining promoters had, more or less, no serious competitors in this style of investment.

The booming days of gold still relatively fresh in memory, they had no great difficulty in persuading large numbers of the public — generally those most distant from the scene of operations — as to the "sterling" worth of the reports of their "experts".

Many of the big goldmining schemes faded out with no other trace than that of the flattened pockets and shattered dreams of the over-hopeful investors. Others left more tangible evidence of their existence — huge holes in the strangest places, heaps of rusting machinery, hundreds of acres of neatly piled gravel and eroded hillsides. For the leavings of others, time and ingenuity found a use — such a one was the Kawarau Dam and bridge.

The "experts" had foretold that when the gates of the dam were closed, the waters of the Kawarau would drain away, exposing the golden hoard which, being beyond the reach of the early miners, still lay at the bottom of that turbulent river just waiting to be picked up.

Practically the whole forty-mile length of the Kawarau's banks were pegged out in chains in anticipation of a rich harvest.

The scene on the day the gates were officially closed was one to remember — the blue waters of Lake Wakatipu surging and boiling past the massive piers and under the gates and bridge of that engineering marvel of the day. The hundreds of excited investors and spectators awaiting the official procedures. The gold coaches and escorts, so thoughtfully provided by the company. The solemn

The Kawarau Falls before the dam / bridge was built. *Photo: Lakes District Museum*

officials and dignitaries keeping themselves well in hand, wearing the importance of this great occasion like a cloak. The marquees, some selling food and gold-pans, others drink and gold-pans, still others just gold-pans, picks and shovels. The talk of the hundreds manning their claims down the river — all awaiting that supreme moment when this example of man's tenacity and skill would overpower Nature and stem this torrent which for countless ages had carried and deposited in its bed an untold wealth of bright yellow gold.

Finally and with due ceremony the great moment had arrived. The gates were wound down and the roar of the Falls was stilled. The waters of the great lake beat in vain against the great iron gates.

With the closing of the gates the crowd surged down to the river, gold-pans at the ready, to follow the receding waters down to the golden hoard. The waters receded — about four feet, and there they stayed. Time would be needed for it to drain away, said the "experts" — but they waited in vain for, fed by its not inconsiderable tributary the Shotover and by the smaller Arrow and Nevis rivers, the old Kawarau maintained a quite respectable flow.

With time the investors, the claim-holders and the "experts", preceded by the promoters, faded quietly from public view. The

The dam/bridge under construction, 1925. Gum trees from the station plantation, killed in the 1923 winter, made for strong scaffolding.

dam, well and faithfully constructed by its engineers, remained, eventually as a bridge to become the major connecting link in one of the main south highways and as a dam to be used in the control of water to the Roxburgh hydro-electric scheme.

The coming of the dam also ended an era for Kawarau Falls Station. Lying to the south of the Kawarau and bounded by Lake Wakatipu on the west, access was only possible by boat or by horseback up the banks of the river or round the shore of the lake. The bulk of the station produce such as wool, grain and sheep was sent out via the lake steamers from a private wharf three miles south of the homestead. Smaller amounts of produce and personnel were usually ferried across the lake in rowboats from the homestead immediately above the Falls.

Although those concerned thought little of this procedure, it had an unnerving effect on anyone experiencing it for the first time — especially if it happened to be at night. A fairly strong current was always running, for here the waters of the lake poured into the beginning of the river — at night the roar of the Falls, less than two hundred yards downstream, seemed to draw steadily closer the further the boat got from the banks. It was relatively easy at night if a lantern could be left on the end of the small wharf: the rower then set the boat at the required angle to the current and

A cargo of rams under tow by the flagship.

all was well — but with no light on a very dark night it was a different story, the rower often ending up in the thick willows above or below the landing — the roar of the Falls usually ensured he try upstream for the landing first.

The fleet of rowboats was three, all wooden and clinker-built; an eight-foot potbellied dinghy, difficult to row, an eighteen-foot beamy and deep-gunwaled boat known as the whaler which was the bulk transporter and was usually towed behind the flagship, a fourteen-footer, a beautifully-built seaworthy boat, very easy to pull.

Many and varied were the loads these boats carried, ranging from drunks, who had to be held down, to waggonettes.

To a boy of eight years these boats were a special joy. He spent a great deal of his time in them, the skill he acquired being barred from use only on very windy rough days. Needless to say he received two sound thrashings, which he still remembers clearly, for contravening orders in this respect — the ominous instruction to follow his father down to the vegetable garden, the clear laying on the line of the relevant facts and then that big plaited belt with the double buckle. No grudge was held, for his father was just, as well as thorough, and firmly believed that if a job was worth doing it was worth doing well.

To the boy, mostly alone in his world of grownups, for no proper schooling was possible — this being carried out by his mother and later a tutor/cowboy (the fortitude the tutor gained at his job was a great help to him in later life) — the days when the team of musterers had to be ferried across were highlights.

The loads of swags and gear, men and dogs (the dogs mostly

swam), the big waggonette, possibly a small mob of rams or other sheep, and the twelve or so horses, all had to be got across.

On arrival, all was assembled near the wharf. The horses, stripped of all but bridle or halter, were tied up in a row, the thirty or forty dogs also tied up, except for one heading dog to hold the small mob of sheep — swags, saddles and harness, tucker baskets, stores, sometimes chaff, there always seemed to be something left to lower into the boats and be ferried across, everyone attending to his allotted task.

Then it was the turn of the waggonette — near the wharf was a specially constructed ramp sloping gradually below the water; this was for loading the waggonettes, gigs, etc. and was also used for the horses' entry. Beside it was a small bank which dropped straight into deep water; this was used for the occasional milk cow or calves that had to be crossed, as well as for loading when the lake was low.

With two men rowing the flagship, the whaler was taken in tow then drawn stern first to the shallow water of the ramp. The waggonette was manhandled backwards astraddle the whaler and the shafts removed and placed inside. It was pushed further into the water, the whaler being kept centrally placed underneath, until it eventually rested on the boat and became waterborne — then, to the steady strokes of the rowers away she went, the wheels revolving and chugging in the water like a paddle-steamer's — it was not an easy row, especially if there was a strong wind.

Now half the men were taken over with their dogs — it was remarkable how fast some of those dogs could swim, the experienced ones treating it as just another job to be done. Some of the younger ones, exhibiting a bad case of nerves and trying to hang on to everything within reach, had to be placed in the boat, to be tossed out when close to the other side as part of their training. Some loved it and would deliberately swim across and back without landing and after a good shake would settle for the final crossing.

Now the boy was ordered well clear, to take up a vantage point for the major show of crossing the horses.

With one man at the oars, usually the Skipper (the author's father), and one in the stern, usually the head shepherd, to lead the horse, the boat was backed into the shallow water. The known good swimmers among the horses were taken first, always leaving one good one behind. The horse was led to the water's edge and a long rein passed to the man in the stern — the long rein of about fourteen feet was necessary when the horse was entering the water as their mode of entry varied. Some would walk quietly out, nuzzling the water, and begin their swim; others, as they felt the water deepen, would make a sudden plunge to get swimming depth

When the waggonette became waterborne its wheels would be "revolving and chugging in the water like a paddle-steamer's."

— the two front feet of a horse in the stern of a small boat in the cold lake water could, literally, be a chilling experience. After the horse was swimming the lead was shortened to about four feet and with the rower keeping just in front and a steady snorting from the horse, the crossing was soon made. Sometimes two good swimmers would be taken at the same time on short leads from either side of the stern.

Then would come the turn of the less experienced and the known jibbers. These were given plenty of time to make their entry, if necessary in the company of an experienced horse led by another boat. If that failed, a tail rope and sterner measures were necessary.

These horses usually provided a nerveracking crossing, demanding all the skill, experience and strength of the rowers for, should a horse break free, he would more than likely go over the Falls. They usually tried to jump into the boat at the start, or would wade so far out then start backing in towing the boat, or, halfway across, would try to turn round, ending up in a panic of plunging and splashing that threatened the safety of the boat. With these particular animals, as the other shore drew near, the rein was let out full length and the boat pulled quickly to the beach, the leader jumping out to lead the horse clear of the boat.

As the horses were landed they were taken over by the waiting men to be dried off with a piece of hoop iron and saddled for their riders.

With all horses over, only the remaining men and dogs were left to ferry across. The excitement of the day had passed, and the boy was free to exercise the patience of his mother or his tutor.

4

SHEEPYARDS AT THE WYE

Against Nature's might
And Nature's pain,
Man's efforts oft are
All in vain.
Yet skill and thought win over fear,
And wisdom gained is ever dear.

THE central point on which all the activities of the station swung was the group of buildings known as the Dip. Here were situated the bunkhouses, cookshop, smithy, stables, etc. A strange name, the Dip, to anyone not familiar with the station's history, for there was no sheep dip or dip of any kind near. The name originated from the days when sheep were washed before shearing, for not far away lay the two lagoons used for the storage of water, and the ditch and tunnel to take it to the hollow used for the purpose.

From the Dip the pack track wound south along the steep flanks of the Remarkables, here falling abruptly into the blue waters of Lake Wakatipu — no gently shelving beaches here, the bluffs plunge straight down into the depths, forming the barrier that halted the first attempts to reach the district by land from the south. Apart from the steamers on the lake, this track was the only direct link with Southland and the railhead at Kingston.

Six miles to the south of the Dip the turbulent waters of Wye Creek add their quota to the broad expanse of Wakatipu.

No mean stream, the Wye — with a catchment area of many thousands of acres, it forms the boundaries of three large blocks and the tiny flats at its mouth form a natural, and the only, gathering place for sheep being mustered in. Here at one time had been a set of sheepyards. They had been placed on the south bank, on the small area of flat ground near the mouth of the creek — to all intents a good situation, sheltered by the high steep bank to the south from the old-man southerlies that funnel up the lake.

It must have been a roaring cold southerly the day they decided on the site, for no hill man would ever easily overlook one of the mountain's basic rules — never camp, never build anything, in the mouth of creek or gully coming from high country.

Today the site is just a tumbled mass of boulders, with no sign that it has ever been otherwise.

It is easy for those who have witnessed such an occurrence to visualise what happened. It was a sultry early spring day after a winter of heavy snow. The big black rainclouds, treading on each others' heels as they came streaming over from the West Coast, were barely able to drag their sagging, pregnant bellies over the lesser peaks. Meeting the high barrier of the Remarkables and unable to rise, they rolled helplessly along the almost vertical faces. Driven by the relentless pressure of the warm west wind and following the path of least resistance, they slid round the point of the Lumber Box spur and went surging up the rapidly ascending Wye Valley.

More and still more came pouring into the gap, like a mob of sheep in a smother, until there was a roiling mass of cloud away up at the head of the valley, that was unable to rise over the divide because of the hundreds of millions of gallons of water weighing them down. There was no escape — the stage was set.

The occasional ominous rumble rapidly increased to an almost continuous, primeval roar, culminating in a thunderclap that seemed to split the sky and under which the very earth trembled — down fell the mass of water and down from the bluffs and chutes slid the untold tons of melting snow.

These avalanches dammed what had been a creek but what was now a raging river. Only briefly would it be denied — gathering speed and volume with every yard it travelled the wall of water, snow and rock, now one hundred and fifty feet deep and over three hundred yards wide and travelling at well over a hundred miles an hour, sliced through the beech forest of the lower valley. Spewing out and down into the very confined and steep last fifteen-hundred-foot drop, it fell like the curse of doom on the peaceful flats below.

In a minute it was all over — in an hour the creek was no more than high, busily cutting a new course through the thousands of tons of rocks, trees and melting snow.

The clouds thinned and a watery sun blinked through — the grey warblers sang again in the tea-tree and Nature settled quietly back to her never-ending work of repair and reclamation.

The inefficient, temporary set of yards which had been erected did not comply with the Skipper's ideas of suitability for handling four or five thousand sheep — so a site was selected, a plan drawn up to match the site and the estimate of materials required compiled and ordered.

In those days the red birch posts came from the Head of the Lake and the timber for yard rails from mills in Southland, all carted on the steamers and landed at the woolshed wharf.

The 12 ft x 6 in x 1 in timber, wire, netting, etc. was packed out with the string of packhorses and the more bulky posts were to

"There was no room for error in judgement or technique by the head packer." These horses were packing timber up the Arrow river gorge, for a back hut on Glencoe station.

be taken down by launch to the beach at Wye Creek, then hauled up to the site on sledges. The string of packhorses with their apparently ungainly loads was not an unusual sight in those days, but it would be an understatement to say that it would be difficult today to gather together a suitable number of experienced pack-horses required for work of this type, let alone the men who could cope with such an operation and the necessary knowledge of what was in reality an art.

There was no room for error in judgment or technique by the head packer, who has final responsibility in such an enterprise. He had to have first-hand knowledge of the idiosyncrasies of the individual horses, for they were as varied in temperament, likes and dislikes as any group of animals, including man.

On one could be hung anything, no matter of what type, size or shape. If the load slipped or got hung up on tree or rock, he would simply stop and patiently wait till matters were set to rights. He was equally as reliable following, or drifting ahead free, as he was confined in the string — a joy to the packer and some consolation for what he had to contend with from some of the other horses.

This one would pack anything just so long as it didn't clink or rattle — if it did, look out! At the first clink or rattle would come a wild-eyed look of consternation followed by a great old performance of dancing sideways away from the noise, pulling back, sudden jumps and nervous trampling until he finished up in a lather of sweat with the string in confusion. Another would give

Awkward loads on a very awkward track. "It would be difficult today to gather together experienced packhorses for work of this type, let alone the men who could cope with such an operation."

his undivided attention to scraping his loads on every available rock, bank or tree, evidencing considerable satisfaction if he could snag a surcingle or pack strap on an obstruction and so display his pulling powers. Others took an almost personal interest in seeing that their loads did not come into contact with anything. The best examples of these would pause at a fully opened twelve-foot gate, deliberate on the space available, edge slowly into the middle, then leap wildly through just in case the posts should suddenly close in. Still another would never settle for less than the leader's place. His opposite number had always to be placed last, for his pet aversion was a lead rope under his tail, even though the crupper was there already. His reaction would be an attempt to create a record for a series of full-powered bucks, all the time clamping firmly on the offending rope with his tail, lest he fail to savour the experience in full and not get at least one of the loads off the hooks.

Then there was the one that had a marked aversion to a high load, such as timber, waving above his head. All would be well until he caught sight of it out of the corner of his eye, then he would go smartly into reverse with the apparent desire to get out from under before it fell on him — it is no wonder packers could display a long-suffering patience as well as a blistering stock of oaths.

However, under the expert supervision, loads were assembled

and nicely balanced, allocated to the different horses and, after several trips, this part of the operation was complete.

Then came the day for the carting of the posts by launch, the boy being highly delighted when told he could come along. The launch selected for the job was the *Thelma*, belonging to and operated by none other than Mr J. Edgar, a well-known personality. Of stocky build and rolling gait, rosy-faced, with a similarly tinted "bugle" of proud proportions, and slow drawling voice, Jock was a character. He operated one of the pioneer tourist enterprises on the lake and had his fingers in many pies. Some of his tourist passengers on long launch trips had every reason to remember the lake and the suggestion that they join in a little game of Crown and Anchor to while away the time — Jock dearly loved his board or a game of poker. These pastimes contributed in no small way to the augmentation of the very reasonable fares — the *Thelma* was a launch that could hold her course unaided for long stretches. Jock had a wide experience and was a skilful and efficient operator of launch — and board.

The *Thelma* was a vessel of some forty feet, of the narrow beam, shallow draft and low freeboard favoured at the long-past period of her design. Interior seating was of the all-round type with several simple cross-seats, passengers being protected from the elements by canvas dodgers, inset with heavy celluloid ports, which let down from around the cabin roof edge. Of uncertain vintage, she was none the less recognised as a sound vessel with good seakeeping qualities. Under the skilled hands of her owner she plied the lake in almost all weathers and is today serving as a fishing boat on the south coast.

The long, curving, white gravel Woolshed beach faces due south into the broad southern arm of the lake. With a fetch of a full eighteen miles a southerly wind soon creates a rough sea and a dangerous lee shore, so the successful conclusion of the operation required a calm day.

It dawned clear and calm. The appointed time for loading was early, as the southerlies tend to get up in the afternoons.

Dawn was just breaking as the Skipper rode away from the homestead, with the boy, all excitement, jogging along behind on the hardy little black pony — three miles to the Dip where they were joined by the shepherds, then on the remaining three miles to the Woolshed.

The *Thelma* came steaming round the point, right on time, as they yarded the horses. Soon all hands were busy trolleying the posts down to the wharf, then passing them down to the launch to be stacked securely inside under the watchful eye of Jock — his repeated exhortations to "mind the varnish" raising many a grin

"A southerly wind will soon create a rough sea and a dangerous lee shore." The long curving woolshed beach.

for the *Thelma* was never in danger of being worn away by spit-and-polish.

As the load of posts grew she sank lower in the water until, with the final posts, less than a foot of freeboard remained — Jock reckoned she couldn't sink anyway, with all that timber aboard, but it was well there was no more than a gentle lap-lap on the white gravel beach.

The *Thelma* left on the four-mile trip to the Wye with the Skipper and the boy aboard, to be dropped off on the return journey. The shepherds, departing on horseback and leading the light draughts for pulling the sledges, were to hasten round to the Wye and meet the launch there.

A perfect day for the job. The snowclad peaks reflected in the only gently heaving blue waters and the *Thelma,* ploughing a much broader furrow than usual, curtsying gently to the swell.

Much too soon for the boy's liking they were rounding to off the mouth of Wye Creek — Jock applied full throttle and drove her firmly onto the soft gravel beach, upon which the shepherds were waiting with the horses and sledges.

A horse and sledge circled, splashing, out then in along each side and unloading began without delay for away down on the lake near Kingston had appeared the thin dark line that foretells the coming of a southerly, and here on this exposed beach there would be no shelter.

All hands worked at top pressure, the boy being detailed to hold a horse during loading. The men soon became soaked with both sweat and water, for all were aware of that fast-advancing line and that the launch must be unloaded and away before the arrival of the wind. Load after load was dragged up through the soft gravel of the beach, to be dumped anyhow above high water. There was no time to carry out the plan to cart the posts direct to the site; that had to go by the board because of the threat from the south.

No remarks about varnish from Jock this time. Gauging the time he needed to get clear of the beach, he kept glancing at the approaching dark wall, which had revealed itself as a real old-man southerly buster. Rolling masses of fog came driving up both sides of the lake, a dark grey mass of cloud swept forward between, blotting out the blue sky and the background hill as it came in a mist of rain. On the water the threatening black line now showed the pale gleam of white horses, indicating the force of the wind. The scene was divided as by a knife, the fast-moving storm wall devouring the peaceful sunlit day as it swept forward.

As the last of the posts were being lifted out came the first puff of wind. The gentle lap of the waters on the beach became an excited splashing and wavelets smacked the side of the *Thelma,* starting her drifting sideways to the beach — Jock leapt to the engine, the boy was yanked aboard and, on full throttle, the now buoyant *Thelma* reversed clear of the beach and dragged her stern up into the teeth of what was by now a full gale. She was safely clear and headed down wind, but now the sun was gone and the roaring wind was driving sheets of spray aboard from the grey waters.

With the waves rapidly increasing in size and the whitecaps hissing past, Jock settled to the task of taking her straight down wind. The Skipper struggled to lower and fasten the wildly flapping canvas dodgers as quickly as he could against the force of the wind.

By now the full force of the storm was upon them, the short-spaced, eight-to ten-foot seas with flying spume from their crests went crashing past — the launch bucked and surged wildly, crashing down into troughs. Occasional waves broke over the stern, bulging in the aft dogers, covering the tough celluloid ports with swirling green water and sending spurts of water in underneath.

Above the roar of the storm, the boy was told in no uncertain terms to get down on the cabin sole and to hold on to one of the cross-seats and stay there. Jock, fighting the helm to keep her dead

Born into a high-country heritage: the author in 1926.

before the seas, was making for the partial shelter of a spur that juts out into the lake a mile north of the Wye. His problem was to work across the seas without broaching to; gradually he worked across, then on full throttle rode a wave close past the point of the bluff and swung into the more subdued waters. Even here, because of the surge and the force of wind, he was unable to round to, but held her, stern on in reverse, with the engine just ticking over — here they remained for an hour while the first fury of the storm gradually eased.

When he judged it safe to proceed Jock took the *Thelma* on, close in to the towering bluffs with the waves smashing along their base, for the breaking effect of the small spur reached well down wind.

As they moved slowly on, the wind continued to drop and the waves moderated — just a typical shortlived southerly buster. By the time the Woolshed was reached, Jock considered it safe to land the Skipper and the boy, so cautiously approaching the shallow end of the beach where, in reverse, stern to the wind and now moderate waves, he held her while the Skipper with the boy in his arms jumped from the bow into the shallows and forged ashore. They turned to watch as the *Thelma* reversed well clear of the beach then, almost rolling her gunwales under, headed for the point around which lay the down-wind run to Queenstown.

They walked along the windy beach to where the horse and pony waited in their yard and were soon on their way back up to the homestead.

With the familiar saddle once more between his knees, slightly skinned though they were from kneeling in the launch, the boy was inclined to chatter — no doubt the reaction from his record silence during the return journey.

5

"THE MORE I SEE OF MEN, THE MORE I SEE IN DOGS" MADAM ROLAND

ABOVE and behind the swelling, tawny-gold breasts of the spurs leading down into the gully, where the hut was situated, the peaks seemed almost to be floating, detached from the earth by the faintly bluish heat haze — their outlines softened and the contrasts between sun and shadow on their few remaining snowfields, muted and pearl-like.

It was a glorious summer's day and a Sunday to boot — a sleepy sun-soaked day, with the faint scent of tussock and matagouri blossom hanging pleasantly on the air. The delicate song of a grey warbler came from further up the gully; barely audible, it blended with and became one with the soft murmur of the creek.

It had become oppressively hot in the corrugated-iron hut where the boy lay reading on his bunk, so he moved outside to the shade of the big elderberry tree where the old shepherd sat oiling a bridle. A couple of little roundbellied pups were romping about, tripping over their own feet and engaging in mock battles. Occasionally one would take hold of the dangling rein of the bridle, sitting back on his haunches in a tug-of-war accompanied by much miniature growling, only to have his tail taken by the other, wanting to join in the fun.

Even at this early age, their hereditary instincts guided their play — the little huntaway yapped and blundered boisterously about while the little heading dog moved with quick darting movements and much "clapping" and "eyeing" and only an occasional yap. It was too hot though for sustained play and they soon flopped down, legs all awry and bellies pressed to the cool earth, their intelligent, joy-filled eyes fixed on the old shepherd.

Here was demonstrated the full strength of the two basic factors that guide the life of the working collie dog — hereditary instinct and, also probably hereditary, desire and liking for human association.

These instincts are strengthened by selective breeding so that the natural instinct is to work in one of three ways — as a huntaway — barking readily and liking to chase stock from behind. Or as a heading dog — seldom if ever barking near stock, inclined to stalk them and hold them in a group. Or as a handy dog — having both instincts in almost equal balance. These instincts are strengthened, controlled and expanded in scope, through tuition and experience.

In some cases a dog may have a particular instinct so highly developed that he is a natural worker, requiring little more in the way of training than that necessary to establish control, the finer working points being gradually built up through experience. Others have to be guided through all phases until these become established with habit.

The training of a working collie is so diverse and wide in scope that a whole book would be required to cover it adequately, so little more than an indication of it can be given here. With a good young dog, training is not so much the establishment of the will of his master to do certain things at a certain command as the establishment of a clear method of communication so that the dog's willingness to serve may be guided to meet the requirements of his master — to bring about a certain result through the cooperative effort of dog and man, the dog being made aware of the intention and the man remaining aware of the difficulties in the way of the dog's attaining it. They should work in instinctive harmony as a team — and often the man is the less efficient member.

As an indication of the mental ability of a good hill dog, it is only necessary to consider that he has to learn from scratch, through the medium of a spoken or whistled command, the intention of his master as well as the requirement to: start running in a particular direction; stop, sit, or stand; bark as required; go faster or slower; run wider or closer to the stock; go gently or give the mob or one particular member a stir up; move below or above; remain in one position; just follow the mob quietly or slew them in a particular direction; hold them there till his master comes; come quickly back when called and remain at heel. In certain cases he has to recognise these requirements from waved signals only.

All these and more are necessary — in general, to become a thinking extension of his master's will. Superimposed on all this basic understanding, according to the individual dog's ability, will come an appreciation of stock sense, even the ability to comprehend the intention of the day's work.

The two prime requirements in the actual training are sympathetic encouragement and clarity of intention — in that order.

One of the great challenges in the actual training is that dogs are as diverse in character and nature as is man himself.

Those people who do not have the intimate daily association of the shepherd with his dogs have little comprehension of the pure mental ability and strength of character that they exhibit in many cases — some are capable of standards of loyalty, devotion and

On Skin-gut Pass, 1934. On country like this the sheepdog must be "a thinking extension of his master's will".

willingness to serve that have humbled many a man. In some, the capacity to reason appears limited only by the physical, not mental, ability to communicate — perhaps this is just as well for man, lest he lose his most devoted friend.

The relationship between man and dog is most capably stated in a line from a poem by Kipling about a pup:

For he has known no other God but thee, since the day that he was born.

"This little feller's grandpappy was a great dog," said the old shepherd, as he bent forward and patted the pup's head. "A really big rough huntaway he was, rough-coated, with feet as big as paddles — Ben I called him. Like most big dogs he had a wonderful nature, never took part in a fight unless some stranger started in on him, and even then he'd try and get out of it by just walking away. Sometimes these scrappers would think old Ben was turning tail and would sail in again and that would be their last mistake; Ben would whip round like a flash, grab the fighter by the small of the back and shake him like a rabbit, then just drop him and walk away — that always put the other one very much in his place.

"The only exception to not joining in a fight I ever saw him make was one day we were climbing out and another musterer's dogs kept picking on one of my young ones. I had broken them up two or three times and told this bloke if he didn't control them I'd wooden them out next time. Old Ben must have realised I was getting pretty annoyed, for when they took to this young dog again and went rolling and worrying down a steep pinch, old Ben took three or four big leaps down the hill, grabbed one by the back and shook hell out of him, grabbed another by the hind leg and nearly tore it off, then chased the other one for about half a mile and gave *him* a doing over.

"By this time this bloke is jumping up and down, spitting sparks and calling old Ben all sorts of a worrier and me worse — he'd just about worked himself up to have a go at me, when Ben comes lolloping back across the hill and strolls slow-like towards this chap. Ben wasn't growling or anything, but a corner of his lip was lifted a bit, showing one of those two-inch canines of his and it was wonderful the way that chap calmed down. He had to do most of the running himself that day, because his dogs couldn't do much else than bark.

"Ben had a tremendous deep bark — it was just like having charges of gelly set off behind you when you barked-up on the hill, and you couldn't work in yards with him very long because you'd end up with a headache. Had to watch him if you were climbing on a young horse; he'd be pottering about behind and with the bit of excitement, might let a *woof* out of him that would scare the daylights out of the horse.

"It was no trouble mustering when old Ben was along — you only had to set him alight at the start and his big bark would go echoing round the bluffs — you would see sheep miles away begin to lift out and draw away. He very seldom had to close in on sheep to get them moving, but if it was necessary and he was asked, he'd go flying in and back them, letting go a few *woofs.* There was never any delay then, they always preferred moving on to getting trampled underfoot and deafened. Occasionally a big wether or a ram would be silly enough to try and make a stand, then old Ben would look back at you with an enquiring sort of look and all you had to do was say 'Hold him' and Ben would stroll up and take hold of him by the nose and give him a gentle shake or two which was always more than enough to get him moving.

"On the beat, Ben was always happier if you put him out above or below to take a beat through on his own — he would move along steadily, giving an occasional bark and gathering in any other mobs, trotting out on to the points of bluffs to make sure all was clear and to check up where you were. He was very gentle with any sheep that was knocking up, and always held back and gave it

A high-country team (Benmore) of seventy years since. Then, as now, the sheepdog was all-important.

plenty of time. Riding round blocks or on the hill, if he spotted a sheep cast or stuck in lawyer he'd go over without being asked and pull it out or try and get it on its feet.

"Although Ben was straight huntaway bred and as full of noise as a drum, he was so brainy he could take the part of a heading dog if necessary. On the hill, the little heading bitch would often get into difficulties pulling a big mob, they might stick up some place and she'd be unable to get them moving; when this happened I feel sure they had a set arrangement between them. She would run up on to a rock or a point and stand there wagging her tail and looking way over to us—often you didn't need to say anything at all to Ben, other times it was only necessary to say 'Give her a hand, Ben', and away he'd go. He would cast out clean as a whistle with never a bark out of him, then come on to the mob from directly behind them and let loose with the noise. That mob would smartly make up its mind to shift — and while this was going on the little bitch would be fizzing about, quite obviously worried that old Ben would overdo things. But he never did. As soon as the mob was moving he'd go quiet again, just tailing quietly after them and letting the little bitch attend to the finer details.

"Sometimes, spot mustering, the heading dog would be fully occupied with one mob, when another would show up heading in the wrong direction then all you needed to say was 'Look at those

The old swim-dip at Kawarau Falls. Modern installations have much simplified this element of sheep-work.

b——s, Ben, they're going to get away on us', and away he'd go, silent as the grave, carving out in the fully approved style, to make a run that any heading dog would have been proud of.

"The most unusual bit of work I ever saw him do was one day when we were dipping. It was one of those old long-swim dips with the narrow chute for sliding them down into the dip, and if it wasn't kept wet the sheep would get enough grip to make a wild leap to try and reach the side of the dip. Very, very few ever did make it, but this day one did.

"When this one made it to the edge and bolted off through the bushes to join the big mob of dipped sheep, the head shepherd said to the bloke on the chute, 'Well, that's one you can catch up and put through when we finish.' I was pushing them up in the forcing pen and didn't pay much attention when I noticed old Ben was not present, thought he might have gone for a drink — then the head shepherd said 'Just look at that!' There was old Ben almost nosing that sheep along that had got away and he brought it right up to the yard; the sheep seemed almost mesmerised with no will of its own. He'd gone down to the mob of about eight hundred dipped sheep, cut this particular one out, then shepherded it back four or five hundred yards to be put through the dip.

"I've seen him staggering on his feet with exhaustion, his feet bleeding, after a particularly bad day's muster — but ask him, and away he would go as if nothing was wrong. It just never entered his head to refuse. I used to keep a pair of dog-boots for his front feet and when you put them on he'd give you a dirty big lick on the face and the old tail would go back up in the air.

"Never a time did I need to lay a hand on him, the whole of his life — if he needed correction when he was younger, all that was necessary was to go crook at him and you'd think his world had collapsed, he would be down in the dumps for hours. But give him a bit of praise or a pat for a job well done and his day was made — his one aim was to serve and he recognised no limit.

"Well," said the old shepherd, "there's a lot more I could tell you about Ben, but it's time these two were back in their pen. You might go and put the billy on. Just remember this though — any shepherd that gets even one dog like Ben in his lifetime can count himself lucky. If he gets two — well, no man deserves it."

THE PASS

A loyal, willing, trusting friend — what greater
Gift, in life, is there for man?
The mutual understanding love that grows — is
Surely part of the Creator's plan.

I asked. As ever willingly you went, though Danger
Opened wide those hungry jaws.
That ledge, that tilted to accursed icy rock,
Destroyed your life — denied your claws.

Old pal, as I knelt beside your poor broken form
And could not hold the tears,
I had vision of a Pass, whose bluffs were Death,
Its shingle chute and rocks — the Years.

Upward to this Pass you ran — against the Rule
That none might enter, but The Dead.
Yet, as you paused to look from there to me,
A kindly Hand reached down — to pat your head.

All must take the deeply worn Track, that leads
From this Green Valley, up the Hill.
Once started on the Way, though pause we would,
Reluctant must we rise — against our will.

Some there are, who more swiftly make the Climb,
Their feet but wings, on known Way.
And some, on hands and knees through Rock and Shingle
Crawl, painfully upward — day by day.

ONE only, has returned to tell, what sunlit Lands
From that high Pass we see.
Yet, that last Climb, more gladly will I make,
If, on those Other Slopes, you wait for me.

6

PACKERS

As jewels or dross
They come and go,
With skill or not
To mix the dough.
Their lot a lonely path do tread.
A saddle home, a tussock, bed.

IN high-country station operations the part played by the packer is of no minor importance — shepherds, like an army, march on their stomachs. The work demands a high output of energy and a day on the hill produces an appetite which encourages disregard for small failings of technique on the part of the packer. Not so on the "days for the King" (wet days in camp), when the packer must really be on his mettle. (Two or more days confined to camp by bad weather and the packer's nerves usually become slightly frayed.)

A good packer is the source of considerable satisfaction all round. The quality of his stews and duffs is held in high regard — nothing unusual being seen in someone stating "I smelt Darbie's stew as I left the top", even though the distance from camp was about three miles. The artistry and ingenuity he could display in the use of the camp oven, when time permitted, in producing such delicacies as fresh damper, scones, "buggers-to-float" (doughnuts) and suchlike were a wonder to behold.

A poor packer is the source of much friction and will get little help with firewood or dishes.

Many packers had reason to be grateful to the owner's wife for the additions to the stores which made his, and all hands', life easier. A good big damp fruitcake made a welcome change from cold duff or "musterer's sponge" (ginger cake), and the addition of ready-mixed duffs and soups ensured the boys had at least a sporting chance of good tucker. From long experience of packers, the housewife usually thoughtfully attached a little note indicating the size of the finished product with further instructions inside.

Peter Rasmussen was a part-time packer who did his best, although the boys had a firmly entrenched belief that his knowledge of the culinary art did not extend past the basic fact that "if you lit a fire and put her on, she got hot".

Over the past few days of this muster, Peter had kept hope high

The author's wife, Jessie S. Jardine, née Reid "ensured the boys had at least a sporting chance of good tucker".

by boasting that when they reached such-and-such a camp they would get a treat, one of "the boss's missus's duffs".

The notable day duly arrived — during the cold meat and spuds Peter fussed importantly round a big billy boiling heartily on its hook. He announced he would wait until all were finished their first course before he lifted the duff so it would be good and hot when served.

Now the moment was at hand — down came the billy and off came the lid. The hut was filled with a most delicious aroma as he fished in the billy with a big fork. Out came great steaming folds of limp cloth in whose centre was a small tight ball about the size of your fist. Holding the steaming mass aloft, Peter stared at it in wonderment, then said in amazement "You wouldn't think that little bugger'd weigh nine pounds."

After untying the string, closer inspection revealed a pale hard mass with bloated currants peeping through; on top reposed a slip of paper, with the now much-faded injunction: *Empty into basin, mix with one pint milk, return to cloth and place in boiling water for three hours.*

The memory of that unrealised duff has haunted the boys down the years. . . .

Three a.m. made a fairly early start to the day's muster, but long experience had taught the head shepherd that on this block an hour in the morning was worth two in the afternoon.

It was the weaning and dipping muster in the height of summer. This block faced due north, only five miles long round the track by the creek, but the four top beats passed through three basins that were specially designed to break a musterer's heart — as one spur was topped the other could be seen a mile away, then as the spur was turned the basin came into sight — huge scollops back into the hill that just went on and on the further you went in. It turned a mile walk into about three miles. There was a temptation to cut straight across, and the man on the third beat fell for it at the first basin — and that was when the day began to go wrong.

It had been one of those fresh crystal-clear mountain mornings as they climbed out from the camp. Just a hint of dew and the sweet smell of tussock, and a whiff or two of aniseed on the cool air — the stars growing paler in the lightening sky. The old hands had remarked on the hint of heat still coming from the ground of the north-facing spurs and predicted a real scorcher.

The young packer had been told to get away before the heat of the day and to get the tents pitched, the tussock cut for bedding and a meal ready for two o'clock at the next camp.

The ewes and lambs drew away readily in the cool morning air, before the cheerful sound of dogs and the occasional long drawn *waa-hoo* from the musterers. All went with a swing — till that first basin.

The head shepherd, a big man, slow of speech but with very definite views, turned a mob of about a hundred and fifty down to reduce their travelling distance. They would normally have been caught by the man below and passed on down to go the short way round on the track.

At this time, however, the man below was halfway across the gully and they passed behind him unseen — the old ewes, sensing the gap, really put their minds to it as they raced back and down, flat out for their old haunts.

A little later the head shepherd moved out on to a point, expecting to see the mob going down the next spur. Instead he saw the man below, now well ahead out of line. With a curse he pictured what had happened. Nothing for it, it would be quicker for himself to go back and catch them. Barking-up so the next man would see him going back, he set off. After some time he caught sight of the tail-end of the string, still going great guns. Out went the old

The packer's skill about to be tested: the start of a muster.

heading dog, running wide and sure. He was holding them, nicely mobbed in a wee basin when the head shepherd caught up. Soon headed in the required direction the old ewes displayed none of the fire and energy of a few minutes before, moving reluctantly ahead with an occasional halfhearted attempt to turn the tables once again.

It was the best part of an hour before the head shepherd got back up to his beat and by now the sun was nearly as hot as his temper.

Hot in the next basin. The last one was a veritable oven. Not a breath of wind, the sun beating down and back from the rocks and tussock. Now that the sheep were camped in small groups in the shade of rock or spur, they no longer drew off for noise but had to be hunted out. The strings moved brokenly as they stopped in small groups, heads down, seeking each others' shade, holding their ground until the huntaways were right on them, before moving on only a short distance. Hard slogging for the dogs, and the men had to cover twice the ground to ensure none were left behind.

It was four o'clock before they reached the last spur and the mob was pouring off in a cloud of dust, to spread out over the flats of the next day's muster.

The satisfaction the head shepherd had in the completion of another day vanished when he saw the campsite far below. No tents, no smoke up, no sign of life. The packer must have gone to the wrong campsite, two miles ahead — glumly they closed on the camp, conjecturing on how he had managed to make the mistake. But no, there were the horses standing hip-shot and head down in the shade of the trees and there, among the tall tussock, lay the pack loads as they had been taken off. Reclining comfortably between two of them lay the packer, sound asleep.

But not for long. With a roar the gang descended on him. He struggled to his feet startled and, although half-drugged with his long sleep in the hot sun, well aware of the enormity of his error.

He was grabbed none too gently by the head shepherd, who looked straight into those now wide-open eyes and said, "You know how we class packers on this station?"

"No."

"Packers, billy-boilers and bastards — and you're third class. Now, GET THAT BLOODY BILLY BOILING!"

7

DEMON OF THE HILLS

A senseless, wild
And ravening beast
From primeval times
Its powers unleashed.
Nature's scourge — to Man the flame
Is servant, master and friend the same.

FIRE — a raging, roaring vortex of unbridled ferocity — tore at the trees, scrub and bracken in the gully, like the fist of some gigantic demented demon. All vegetation within a quarter of a mile was laid flat by the hurricane force of the whirling gale of air being sucked in to feed its lust. The screech of trees bursting apart came clearly above the deafening roar.

There was little smoke — but a whirling pillar of fire, a quarter of a mile across, shot straight up in the air. Four hundred feet up, great gobs of still flaming gas were carried away by the natural wind. Still-burning branches and bunches of bracken spewed out of the top of the pillar, the small branches falling like a shower of incendiary bombs on the surrounding scrub and the lighter parts carried farther out to start a hundred new fires, all rushing back in to feed the giant.

Perhaps it lasted five minutes. Then, consumed by its own insensate fury, it suddenly collapsed. Then it began burning again on a front, racing up the gully as if to savour once more, in some suitable spot, the fullness of its power.

On the bracken slopes on either side of the gully, walls of flame fifty feet high moved onward with a hissing roar. Fed by their own suction and the steady wind, they sent burning gases and fronds racing ahead, compounding the speed of advance. Here and there birds flew up and away, some tumbling down as if shot, caught by a gust of superheated air — no man or animal caught above this wall of flame could hope to outdistance it. It was impossible, because of the intense heat, even to face the fire from a distance of a hundred yards upwind.

Above the bracken line lay the sweeping face of the hill — silver tussock and danthonia, still dry from the winter's frosts — and here there would be sheep. Four thousand grazed this block and the fire had started right in the middle. The Staircase block was well alight.

Demon of the hills.

One of the few things that will stop a high-country man's heart in mid-beat is the word *fire* — especially in a dry spell. The work of years can be wiped away in a day. Anyone who has had to stand by helplessly and watch the destruction of sheep by fire, who has had to slaughter those still alive but blind and burned almost beyond recognition, or who has fought a hill fire for almost twenty-four hours at a stretch without real rest — he'll have experienced a nightmare from which he will never again be entirely free.

The work of firefighting, swinging heavy water-soaked sacks to beat out the flames on a steep hill slope, combined with the heat, smoke and flying ashes, brings a man to the point of exhaustion quicker than most other jobs — men have stumbled down in the tussocks and dropped fast asleep before they would give in.

In the past the only means that could be used in fighting a bad hill fire were experience, manpower and organisation. Today, where water is available, the helicopter with a monsoon bucket is a godsend, yet man's intellect and willpower must remain the major factors in subduing any fire.

Firefighting is just that — it has some similarities to an actual battle, in that hope of success will depend on a carefully assessed strategy under the sole direction of one man. If he is wise he will seek and listen to opinions, which he will weigh against his own experience.

A high-country team of 1926. The basic requirements of the job have not changed: toughness, energy and common sense. *Back row:* D. McKenzie, V. Baker, W. Miller, L. 'Bluey' Oakes (Packer). *Front row:* A. Chapman-Cohen, D. Rowe (Head Shepherd), Dickson Jardine, G. Metzger, S. Stevenson.

His will be the decision, whether to fight or to accept the results the fire may bring, and many factors will enter into this decision. Among these will be present and possible weather, terrain, natural boundaries that would limit the fire, manpower, and any other assistance that may be available.

Once the decision to fight is made and a plan of attack decided on, it is essential that all concerned are briefed and clearly understand what is intended. Coordination and control of the firefighters is absolutely essential, as much for their own safety as for the success of the attack. It is notoriously difficult to maintain contact with and control a large number of men, perhaps hundreds, split into groups and often widely separated, more especially so at night. Confusion is to be avoided at all costs — this can only be accomplished by dividing the available force into sections of about twelve men, each under the control of a leader, who should have a list of their names. The leader will be responsible for on-the-spot decisions but must not diverge from the instructions given him by the central controller. As a swift withdrawal may be necessary because of a change of wind or for some other reason, a system should be evolved and clearly understood. No man or group should withdraw without advising his neighbour. A particular location should be selected by the controller as his base, so that all know where to contact him. He should retain two or three experienced men to act as messengers, and a couple of walkie-talkies are a big help.

The controller has to arrange food and water supply to the fighters and relief teams, also patrols for the areas brought under control. He has a big job and a heavy responsibility.

This day the first tell-tale billows of smoke to go rolling up the gentle face of the Staircase had fortunately been noticed by one of the boys, ten miles away back at the station. He dropped what he was doing and galloped straight back to the yards where the others were drafting. The head shepherd sent him on to tell the Skipper and, taking time only to throw open the yard gates, set all hands to collecting old bags.

Within half an hour the Skipper had arrived and taken charge. His first consideration was to see whether the fire could be "nipped in the bud" — so, with dogs reduced to two per man and an ample supply of sacks, they set off. The scene that met their gaze when they arrived was what has already been detailed — it was going to be a big job.

As the wind was blowing at only a small angle to the slope, the head shepherd and the young shepherd were sent to climb fast and wide of the windward side of the fire. Their job was, if possible, to outdistance it and set any sheep directly above moving out of immediate danger.

Messages for assistance were sent to the engineer in charge of the Ministry of Works camp and to the homestead for food and a cook, slashers and shovels, and more sacks.

In less than an hour seventy men had gathered and the plan of campaign had been spelt out — the aim was to contain the fire from the sides, from the bottom up, working towards a final attack in the lighter tussock of the tops, at about two or three the next morning when tussock fires are generally at their weakest.

As it was not possible to fight the fire directly in the six-foot-high bracken, it would have to be back-burned from specially cleared breaks well out on each side. This was done from the top edge of the bracken back down, a section at a time, so control could be maintained. A short burn was also put in in the tussock towards the edge of the advancing fire. Back-burning is desperate work; make one error — and all that results will be an enlargement of the fire.

With the bracken breaks established, and back-burning nicely at no greater cost than the depilation of many hairy arms and eyebrows, the stage was set for the main attack. In the gathering darkness sections of men fought their way up the edge of the fire. As they subdued their section they leapfrogged above the other sections to a fresh attack. Often forced back by billowing clouds of smoke carried by a slight shift in the wind, forced to draw their breath for minutes at a time from the relatively clear air at ground level, often forced to fight with frantic energy to contain sections where the tussock and grass flared savagely as if soaked with kerosene, yet the dark that signified extinction moved with remarkable speed up the flanks of the fire.

By midnight, with the fire now losing its savage daytime intensity, the teams from each side were converging on the final tongue of flames, in the scattered snow-tussock and lighter grasses at just under five thousand feet. Suddenly darkness descended on the hill. The fire had been beaten.

In the void of dark below them, like sparkling stars in an inverted sky, thousands of points of light glowed briefly to life where the smouldering remains of tussock or scrub were briefly fanned by the dying wind. With blackened faces and bloodshot eyes, semi-blind from constant staring from blazing tussock to black darkness, dog-tired but happy with their success, the firefighters stumbled down through the darkness to the waiting food, tea and beer — firefighting raises a thirst which is hard to quench.

The young shepherd and another were left to keep vigil throughout the remainder of the night to stamp out any small outbreak or raise the alarm if any resurgence got out of hand. Few things are as disappointing as, having successfully defeated a fire, to have to cope with a fresh outbreak in the heat of another day.

Daylight revealed some three thousand acres blackened, with sheep losses of a small mob of fifteen which had been caught in the heavy bracken. Experience, determination, willing manpower and luck with the weather had averted a real disaster.

There have been many other bigger fires over the years throughout the high country — the Skipper had to contend with one on Glencoe Station that burned for five days in midsummer, blackening many thousands of acres. Three hundred men fought that blaze.

Most accidental fires are due to downright carelessness and lack of appreciation of what is involved — the Glencoe fire was due to trampers walking off and leaving a campfire unquenched. This Staircase fire had been due to the same cause.

The trampers had boiled their billy beside the chuckling stream in the gully, where the beech trees cast their welcome shade over the ribbonwood, fuschia and young lancewoods standing tall in the bracken. Now only charred and blackened trunks remained. No birds sang. It would take Nature a hundred years or more to repair the damage.

8

CROSSING THE WYE

Now Youth and Love
As ever, run
Hand in hand
With manhood won.
For life is good and wide and free,
And, all in all, is all for thee.

SHEETS of spray from the falls above him shot past on the savagely displaced gusts of air — the roaring surge of water between and around the rocks and the buffeting of the wind and spray made thinking a conscious effort.

The gap between the two huge boulders was only about five feet, the one on which he stood being slightly higher than the other — an easy jump and no trouble to the dogs. But between and around those boulders poured a millrace of water of almost primeval power and ferocity as it tore down into the black boulder-strewn gorge below. One mistake here and ——!

He looked back up the steep shadowed face behind, covered with broadleaf, scrub and vine which hid the shattered rocks and miniature bluffs of the old rock fall down which he had come; he thought of the climb back up, then down a thousand feet to the crossing, and the longer climb up the opposite sunlit face — by the time he'd done all that, the small mob stringing round high up on the opposite face would be well gone and it might take hours to catch them.

His assessment from the ridge above had been correct. A crossing *could* be made here — but that assessment had been made high up in the sun of the open tussock: here in the sunless black gorge with the oppressive power of the water dominating everything it assumed a different dimension. Even the dogs were uneasy, half crouched with claws bared to the slippery black rock.

"You know your judgement is correct; it is an easy jump," he told himself. "Shut out the noise, disregard the rushing water, concentrate," — two steps back and — he felt the jar of solid rock of the other side come up through his boots with a sense of relief out of all proportion to the actual accomplishment.

Two of the dogs, Roy and Stop, had followed without question, as they always did. But Lass, that wild red-eyed beardie bitch was

Dogs, hillstick, tuckerbag — and a load of responsibility. The author in 1936.

still skittering about, all indecision, on the other side.

She had obviously sensed her master's initial indecision and had panicked. He tried to talk to her, but she could not hear him above the noise — she made a scratching wild leap that carried her right past, over the edge of the rock into the surging water on the top side, where she was drawn under. He rushed to the bottom side to try and grab her as she came out — only the surging water, had she gone already? For a few seconds he stared in unbelief, then tore back to the top side — not a sign of her there. Then suddenly it came to him — she must be pinned. Flopping down on his stomach he looked underneath and there she was, about two feet under, flattened against the rock by the pressure, she stared up through the crystal-clear water — a quick dive with an arm, a hefty pull and she lay a bedraggled heap on the rock.

Lifting her by the hind legs let loose a torrent of water from inside, a few swings and some chest-pummelling brought rapidly strengthening gasps — five minutes later and with many a shake of her thick, wiry coat, she was her usual intractable self.

They had struck a very hot dry spell for the dipping muster on Glencoe Station. This day as they returned to the homestead it had become more humid and close, the sun blazing down into the gullies with no breeze to break the oppressive heat, although some relief came from the slowly drifting shadows of the towering cumulus clouds which were marching up out of the west.

They had been mustering steadily for some time, with no "days for the King", and all welcomed the return to the homestead with the chance of a break — some to return to their homes for the weekend, others bent on a short session at one of the pubs. The boy decided to ride down to Arrowtown and meet his girlfriend at the "pictures" — quite by chance, of course! After tea, during which he had to endure the usual chiacking from the other members of the team he caught and saddled his horse and set out down the bridle and Tobin's Track, which sidled down the steep face of the Terrace to Arrowtown, 1100 feet below.

As he topped the ridge the whole of the Wakatipu area unfolded below — the green farmlands, ringed by the mountain peaks with, in the west, the great stretch of Lake Wakatipu, a dull silver now, backed by its enclosing mountains, which were in dark silhouette under the blue-black base of the slowly roiling mass of cloud to the west. Undeterred by the thought that soon those clouds would bring a downpour, and thankful his oilskins were still on the saddle, he turned down the track. His horse, used to mountain tracks and fit after a long spell, attended strictly to business and with puffs of the fine mica dust rising at every hoof-fall dropped steadily down towards the township.

The film was *Daddy Longlegs* — silent, of course, but supported by the piano accompanist who had long since mastered the technique of accentuating the unfolding story and moods of the celluloid sagas. Lengthened by a couple of projector breakdowns it was enjoyed by all, especially those young ones holding hands in the dark.

With the show and the reluctant parting eventually over, the boy collected his horse from its stall in the big old coaching stables and set off back across the river and up the hill.

The night was still oppressively hot and still, completely dark under the heavy cloud mass. Quick glimpses of the trail were given by the pale intermittent flicker of sheet lightning in the west.

Although the rider could see nothing his horse plodded confidently on and could be relied on to keep to the track.

They were nearing the top of the face when the boy was amazed and intrigued to see two small blue flames, about the size of candle flames, gradually appear from the tips of his horse's ears. The horse plodded on in apparent unconcern, with the whole top of his head now illuminated by the eerie blue light from the little flames, flicking back and forward with the horse's ears.

Fortunately for the boy's peace of mind, he had read of this effect — St Elmo's Fire — and was able to enjoy this display of one of Nature's little tricks.

The storm was now rapidly moving in from the west. The crash and rumble of the thunder was almost continuous, great forks of lightning sliced through the clouds close overhead, the crash of the accompanying thunder was almost a physical blow. With the first few drops of rain the little blue lights faded and vanished.

Their way lighted now by the almost continuous lightning flashes, they pressed on at a trot, the horse as anxious as his rider to get home before the threatened downpour.

As they turned into the stable yard the heavens fairly opened. The boy unsaddled in the secure darkness of the stable, to the roar of rain on the iron roof.

9

ON THE BOUNDARY

If man would
Close to Nature be,
Let him take a
Job on the boundary.
Alone with dogs and hills and sheep,
His own company he'll learn to keep.

FOR those who have not enjoyed the doubtful pleasures of keeping a boundary, an explanation would be in order. The pattern of operations on a high-country run is dictated by the inexorable cycle of the seasons. In their allotted place within the slowly rotating cycle, the seasons come and go in steady transformation, one to the other. This is Nature's plan, her rule; but, as in all things, it is the exception that proves the rule.

Sometimes, with a dramatic and almost unbelievable abruptness, the land which is one day drowsing in the heat of an Indian summer's day, can the next be lying silent under a shroud of snow in conditions as severe as those of arctic lands. Because of the time of year at which these sudden falls occur, the high lands almost invariably remain snowbound, held fast in winter's tightening grip until such time as the cycle releases them once more to life.

It is the recurring threat of these exceptions, even though summer still apparently rules the land, that dictates the timing of the fall muster when, in accord with the past experience of the runholder, it is no longer safe to leave the sheep out on the really high country. They are brought down to the station, the ewes drafted off for crutching and mating, and the wethers eyeclipped if this is required. The ewes go out to their tupping blocks to await the rams, the wethers and possibly the hoggets are driven out to the winter block where they remain for the winter.

Though the spot muster has yet to be done, everyone feels the tension relax — the flock is safe.

The basic requirements in the selection of a winter block are: that it should have sufficient scope of country below the settled winter snowline to carry the necessary number of sheep for the required period; that its aspect should be such that snowfalls reaching lower levels have every chance of clearing back to the normal snowline. In high-country terms it should be "safe" country — that is, as safe as the majority of the high country can

"Aerial oversowing and top-dressing have brought a new era to the high country."

ever be. (The use of this term, in these instances, is only relative.)

On the average high-country run, the area of land within each individual lease that is suited to the above requirements dictates the numbers of stock that the property can carry—a well-balanced property is one on which the summer carrying capacity on the high lands is equalled by its wintering ability on the lower country.

The early surveyors of the Lands Department, in their laying down of the boundaries of the leases, often lacked the specialised knowledge required to establish a balanced unit. Added to this, in many areas, in many instances, there was no local experience available to aid them in their decision. The result was that a great number of unbalanced leases were established; with the passing of the years the Department has sought to establish a more uniform balance within leases by the reallocation of areas or the combining of leases. It has been a slow process, founded on the bitter experience of those high-country men who, seeking to make a go of the unbalanced lease, had every reason to curse the lack of knowledge of the original surveyors or Department officers.

The proper pastoral management required in the use of native grasslands in their original state is vastly different from that used with pastures composed of improved species: the native grasses will not stand continuous grazing; they must be spelled. Where sufficient scope of land is available and natural boundaries exist, the rotational, full season's spelling of blocks yields beneficial results.

The winter blocks are always kept free of stock during the summer months to allow the cover to make maximum growth — this, though dried by the summer sun, is still palatable and provides the winter's supply of feed.

The introduction of aerial oversowing and topdressing of the lower mountain lands has brought a new era to the high country. The introduction of the improved grasses has increased the carrying capacity by a considerable extent, more especially on the winter blocks where the longer growing period of the improved strains and their ability to withstand intensive grazing is a great asset.

In the past, few indeed were the high-country runs fortunate enough to have within their bounds a safe winter block isolated from the main ranges by natural or fenced boundaries. More generally these blocks were no more than a great stretch of the more favoured part of the main range. Where this situation applied it was imperative that the sheep be confined, not only to the particular block, but below the snowline. This duty became the responsibility of a shepherd, the boundary-keeper. The practice was one born of necessity and is unique to the high-country pastoral industry of New Zealand.

Today the situation is greatly altered. The majority of runs have snowline fences on their winter blocks, with access tracks leading out to the fenceline to aid in the still prudent hunting-down when the weather threatens. On the few remaining runs where a snowline fence is impractical or uneconomic, boundary-keeping remains the only method of ensuring the safety of the flock — it is to these properties and to the past, that the following brief description of the life of a boundary-keeper will apply.

With the release of the sheep to their winter range the work of the boundary-keeper begins — like dropped mercury they will spread, relishing the untouched grasslands and their release from the confines of paddock and yard.

Despite the gastronomic attractions of the fresh feed, they are drawn to seek again the high basins in which they have spent the summer, whence they have been so recently and forcibly removed. They lose little time in seeking to make their return. The high passes that would lead them back are ever-beckoning, and no snow has yet fallen to bar their way.

Climb they do, and fast — the only bar to their designs the vigilance of a man and the dominating force of his dogs.

They must be hunted back down, day after day, as they persistently probe for a way back to their favourite haunts — they must be kept below the level of the settled winter snowline lest they be trapped by a sudden fall, and woe betide the boundary-keeper who allows sheep past and involves his fellows in a bout of snow-raking.

Days, even weeks, pass before the scattered mob accept to some degree that their way is barred and reluctantly settle down on the block — some never do, they keep trying until snow eventually and finally seals them in.

On all but the very worst of days the boundary-keeper must climb out on to his beat to ensure they remain below the snowline; even when the mob is settled there are always those persistent leaders seeking to draw a string of their willing followers back to the forbidden lands — a change in wind or weather can also upset the whole mob, setting them drifting restlessly over the length of the block.

Clear moonlit nights are bad, for then the persistent ones can climb all night — after a bright moonlit night it always pays to take more country within the sweep of the beat, but even then one can come on the telltale tracks of a mob winding up through the big snow tussocks, heading out and away to the distant pass. They must be immediately tracked down, caught and hunted back down to the lower spurs for, once on their way, they will not pause even to graze until their favourite haunts are once more in sight. Should they get clear away to reach their goal without being noticed, certain death awaits them in the coming snows.

As day follows day, the boundary-keeper, more often than not camped alone at a back hut, falls into a routine. Though the same job has to be done each day, no one day is the same as another — there are the busy days, especially when the boundary is first started, and days of sun and peace when all is well and the beat is no more than a pleasant day's walk; anxious days, when gathering stormclouds threaten a snowfall and the boundary must be hunted well down; days when a raging storm strikes before he can get off the hill; and wet "days for the King" spent at ease in the hut, with plenty of time to make a fancy stew and do a bit of washing with no worries over the sheep, for even the big wethers are no more keen than anyone else to push their way up through the soaking wet snow tussocks.

The boundary-keeper is his own boss, but is always conscious that he bears the sole responsibility for the safety of the many thousands of sheep in his charge.

It is a position of trust, a tribute to his ability and skill of which any man can be proud.

To the youthful shepherd the orders to take over the boundary were a great thrill. It was the first official recognition of his ability, and he settled in to his solitary camp at the Wye hut in high spirits as the other shepherds returned to the station after liberating the sheep. He had assisted on the boundary before and knew what was required.

In surroundings like these the boundary-keeper lived while he kept his lonely vigil. The Staircase hut, 1930.

With his day starting at daybreak, the length of a boundary-keeper's day depends on several factors, mainly the extent and length of his boundary. This one involved a daily climb of just under 5000 feet with a distance out and back of fourteen miles, the actual time involved depending on the work encountered, usually between six and seven hours.

Hardened and fit as he was from a full season on the hill, the actual climbing and walking were of little concern. When so inclined, he sought to reduce his best time for the actual climb out, just as an athlete may seek to reduce his time over a particular distance.

If he had planned extra chores for the day, such as killing mutton or dogtucker, or if it was the day for the weekly visit of the Skipper with fresh bread, papers, mail, and perhaps a big pie and a more than generous-sized fruitcake, no time would be spent with the fieldglasses studying the disposition of the widely scattered mob grazing the gentle slopes far below, or the sweep of lake and serried mountain tops. The huntaway pup would be left at home on these days as well, for there could be no time spent in his coaching or in correcting his inevitable mistakes. When left behind, even in the company of older dogs, his mournful wails could be heard long after the hut had passed from view. The boundary is a grand place for breaking in a young huntaway, providing straight hunts far down the spurs, with fit, fast-moving wethers well able to outdistance an over-zealous pup.

With the passing days he came to know like the back of his hand the route taken in the climb out and his beat on the hill. He became well acquainted with the gullies and spurs up which the restless ones sought to escape; the high points from which the greatest scope of country would come under his eye; he had set points at which he stopped for a spell and a bite of lunch; he began to recognise the persistent tryers individually and his suspicions would be raised if he failed to sight them during the course of the day.

Though his chances to talk with others were few, he had the changing interests of each day and the companionship of the dogs — and always he watched the weather glass in the hope that snow would come and allow him to complete the job without having to pass it on to someone else to finish. The rugby season would be starting soon and he had no wish to be saddled with the boundary when that important part of the year's activities was taking place.

Sometimes, with an open winter, the boundary would drag on and on, to become an irritating job for the shepherds and a continuing source of worry to the Skipper and the head shepherd.

He had been wakened, not by the clatter of the loudly ticking old alarm clock, but by the moan of the wind round the corner of the hut. He lay in the dark, drowsily pondering whether to strike a match and check the time when, just as he was drifting back to sleep, it came to him that, somehow, this wind had a different sound and that the air too had a different smell. Now wide awake, he hastily struck a match and, lighting the candle, went to look at the barometer. The "bottom had fallen out of it", it had dropped more than ten points overnight. It had remained steadily low for the past three days and now, with this rapid fall, he knew snow was upon him.

He checked the time: a bit after three o'clock. Hell — four hours to daylight. He stirred up the embers of last night's fire, tossing on the kindling and wood ready to hand. As it burst into crackling life the small camp oven and billy of water were slid into place. By the time he had washed and dressed, the chops were ready to turn and the billy was humming.

As he set about cutting a lunch he retravelled the boundary in his thoughts, reviewing the situation as he had left it the previous day. Yesterday the sheep had been restless: where, normally, they would have been spread fairly evenly over the whole block in small groups of a dozen or so, grazing in open formation as they moved slowly up wind or resting on the sunny side of the spurs, some lying down and the rest standing about viewing the scene and contentedly chewing the cud — yesterday they had formed into small mobs of a hundred or so, grazing in more or less compact formation as they moved jerkily forward across the face of a spur. Suddenly, for no apparent reason, they would begin to string off rapidly in single file to the next spur, where the process was repeated. Well down on the face he had seen two strings of upward of five hundred sheep, moving as if they were being driven; this was the more unusual because the two mobs were stringing in opposite directions and passing each other by only a few hundred yards.

Halfway through the beat he had topped a spur to come face to face with one of the well-known "pokers", a big dark wether with one twisted horn. Behind this sheep came a string of fifty or more, all busily winding up the spur. With a disgusted snort the big leader had wheeled and bolted back down the hill, the sharp about-face leaving some of the string gazing bemusedly after him, their now rapidly descending leader.

The young shepherd had given the whole boundary a good tuck in, just for luck. Now, thanking his lucky stars that he had done so, he realised that if a bad fall was in the offing they would have to be tightened down still further, so he settled to breakfast.

Breakfast was usually an unhurried meal, eaten leisurely as he read a part of a more recent paper. This morning he ate methodically, and without pause stowing away a hearty meal, for this could be a long day.

Wondering just how much ground he could cover before the storm struck, he simply stacked the dishes and doused the fire. Outside in the boisterous dark he fumbled for the dogs' chains; he would take the full team, excepting the pup, for there could be

Snow. The boundary-keeper's job is to prevent sheep straying up to the higher slopes, where they may be trapped by the first heavy snowfall.

"He had set points at which he stopped for a spell and a bite of lunch . . ." and a read of his two-weeks-old lunch paper.

much to be done. As he climbed away from the hut the pup's wails were soon lost in the gusts of wind.

The ability to travel over hill country in the dark depends on three factors: an intimate knowledge of the route to be travelled; experience; and concentration. If one or another of these is missing, progress can be made only by resort to the use of a light, which tends to reduce appreciation of position and angle of climb or descent; as a pilot flies blind by map and instruments, so a hill man climbs in the dark, following a mental map of his route with reference to the feel of the terrain under his feet. One advantage the shepherd has over the pilot is that any error is brought home to him almost immediately.

He climbed steadily upward, making good time in sections where the old heading dog moved out in front. The first grey light of dawn was just beginning to filter through the sullen clouds scudding past close overhead as he reached the highest point of the climb. Here, as he waited for the light to strengthen, he took shelter, for the wind had a bite to it now — the tall snow tussock, combed flat with the wind, whiplashed in the eddies round the big rock.

Just as soon as it was light enough to see, he moved on down the boundary, travelling fast from high point to high point; at each of these he halted and, as the wind buffeted and tore at his clothes and sought to suck his breath away, he searched, through streaming eyes, the slopes below for sight of sheep.

He had reached the last basin, the turning point on the boundary, before he struck them. Here he usually stopped for lunch but today did not pause, for the peaks far to the south and west were disappearing, one by one, beneath a grey pall.

He dropped down the hill to the level of the return beat and started back — now he began to strike more sheep and these he hunted well down. In this wind the dogs could not hear whistled or spoken commands and he worked them by arm signals. When they had hunted the sheep far enough down, as they looked back for further signals, he would turn and walk on, the dogs starting back up the hill to pick up his scent and catch him up further on.

It was slow going as he worked his way back, for all sheep within reach were hunted down.

He had got more than halfway back when the first patches of fog began to sweep across the face. As he moved on, working as fast as he could, the light began to fail, making it difficult to see any distance. The wind too, had altered, it now varied from fierce gusts to dead calm. During the latter the rumbling roar of the wind through the passes high above could be heard quite plainly.

As he reached the final spur of the block the wind suddenly swung full to the south and blew with steady force. He stopped and looked back along the big face. Only a couple of miles behind him a fast-advancing grey wall blotted out all as it came — he turned and ran down the spur, heading for the pack track skirting the edge of the lake. As he reached the bottom and knelt for a drink at the creek, he suddenly realised he had had nothing to eat all day, so he sank back into the bracken out of the wind and unwrapped his lunch.

A flurry of fine snow went dancing through the gully. The wind seemed to ease, just a little, and then thick snow began to fall — big flakes, more than an inch across, drifting silently down to settle in his hair and on his lunch.

He munched on in high content. He had done all that could be done, there would be a boundary, and — there was the match coming up against Queenstown next Saturday.

10

SHADOWS ON THE HILL

Conceited man is
Far from wise,
Who dreams the myth
That wisdom lies
In puny effort and known ways,
When dark the glass and dim his gaze.

MOUNTAINS — withdrawn, apart. Since the dawn of time they have been a focal point for human emotions — of inspiration or despair, love or fear, both a challenge and a barrier. The setting in which man down the years has known great personal triumph or bitter defeat. The abode of spirits, both good and bad.

To those who journey into the realm of the mountain gods, death stalks as attentively behind as does joy go ahead, extolling the beauties and whispering of what lies just beyond that top or spur.

The mountains have been selected as the setting for many of the great moments in recorded history and form the central theme in many of our great art and literary treasures. They are loved with a deep and abiding love by the people who shelter in their arms and are longed for by those kindred spirits who are forced to live in exile in the cities or plains; those whom the mountains have taken as their own can know no peace apart.

Mark well — the mountains are in very fact a "different world", one which welcomes and nurtures yet exacts full toll for disrespect. Man is tolerated, yet is made aware he is but a tiny crystal in the great chain of Nature — only his spirit is free to rise above all claims the mountain can make.

In the high mountains man senses he is near to some inexplicable flow of power — is this but the stimulus of strange surroundings, or the body's almost unperceived reaction to the world's venting back to space, through its highest points, the excess of its colossal store of negative electricity? Or are these mountains indeed "the footstools of the Lord" and does man dimly sense a flow of cosmic power, as yet unknown? A life-force, part-physical, part-spiritual (for who shall say where the boundary lies?), which flows most strongly into the earth at its highest points. Can it be that without this life-force all would wither and die away, to become "an abomination of desolation'' — whether the flow is in or out, whatever it is, it is there.

The highest peaks have been physically conquered, yet, as long as man may live, will he seek the spirit of the hills.

Mountain peoples, of all races, have woven a tapestry of legend around their beloved hills — a realm of mystery and a source of ever-changing beauty.

Who has not been enchanted, on some bright summer's day, by the sight of small cloud-shadows gliding effortlessly across a sunlit mountainside? Those who know and love the mountains can sympathise with the strongly held belief of the natives of central New Guinea that the spirits of their dead become — ''shadows on the hill''.

With only three more days of spot mustering left to do, the three men were to camp that night at the small stone Two Mile hut, situated in a basin at over 6000 feet, closely guarded by the sheer rock of the surrounding peaks.

It was just on dark, on one of those completely still and clear winter nights, as they crossed the saddle into the basin and made for the hut. The first snowfall of winter had come, fortunately only six inches, which now was frozen crisply over the huge scattered boulders, scree and tussock — a foundation for the heavier falls yet to come. Above, the stars glittered behind the jagged peaks, the black of the plunging bluffs and scattered boulders contrasting sharply with the stark white of the snow. Over all lay the living silence peculiar to high places. The temperature had dropped to the point where it hurt one's throat to breathe, and they hurried on to the shelter of the hut, down through the ice-clad tussocks with the snow crunching and squeaking under their feet.

The hut, built by one of the boundary-keepers of times past, twelve feet by eight, with low, dry stone walls packed with mud, a squat stone chimney and a high-pitched corrugated-iron roof and one tiny window — was built, literally, into the small spur. Secure from avalanche it appeared to crouch back into the ground for protection from the cold and the raging blizzards that sweep these high lands.

Inside, the field bunk of planks covered with dry snow tussock took up half the room. In the remaining free space of dirt floor paved with stones was the large tin-lined box which served the dual purpose of small table and storage for a few emergency rations, secure against the attentions of the field mice who, for reasons best known to themselves, prefer to carry through their existence amidst the privations of the high country. A few pots

A dog's life, high-country style. So long as he is adequately fed he's content to spend the night in a snowdrift.

and billies, a heap of peats and a little coal and wood and that was all.

Small heaps of snow lay inside the door and in the fireplace. A fine sprinkle powdered the blankets hung from a rafter below the snowgrass-and-bag-lined roof — the fine dry snow of these altitudes, driven by the weight of wind, will find the smallest chink and filter through.

The cold and damp of lack of use hung heavily inside, yet here could be found warmth, rest and, not least, a good big billy of steaming hot tea to end a long hard day.

They soon had a couple of candles and a good fire burning, for this hut would not be used till next season and they could afford to be heavy on the fuel — presently, with a hot meal put away and the hut alive with the deep-red glow of the peat blocks burning, they started washing up before turning in.

Outside, it had not taken the dogs long to burrow well in under the old-man snow tussocks, into the dead dry fronds of past seasons — here they would remain quietly till morning, reasonably warm and protected from the penetrating cold.

They were just finishing the washing up when they heard, faintly, the sound of someone coming down the spur towards the hut. They knew at the homestead that the men would be camped here tonight, so it must be some very serious and urgent reason that would bring someone this distance out on to the hill, in these conditions, to finish his climb in the dark.

Silently and in some consternation they waited as the footsteps drew steadily closer, the crunch of the snow and the tinkle of the ice brushed from the snow tussock growing louder as he approached the hut. There came a sudden burst of barking and snarling from the dogs and the men waited for the door to be opened. The footsteps did not stop but carried on past the hut down towards the creek.

Only momentarily stilled in amazement they threw open the door. Clearly lit by the just risen moon, the basin lay stark and white. The dogs, with hackles up, stood staring towards the creek, which the men quite clearly heard him jump, yet they could see no one. They called out, but their only answer was the echo from the bluffs across the creek and the sound of the footsteps receding as he wound his way through the boulders in the deep shadow below the bluffs. Then all was still.

Quickly they pulled on their boots to catch up to him, reaching the creek before they realised *there were no tracks in the snow,* on either side, or below the bluffs. Above the hut on the spur the snow lay undisturbed as it had fallen. Baffled, they searched and then, as the dogs retreated one by one to their tussocks and the biting cold tightened its grip, they returned uneasily to the hut — to the security of its warmth and light, closing the door on that harsh mountain world they thought they knew so well, but which had now revealed a dimension that stilled their conversation and sent them thoughtfully to bed.

Ice on a snow tussock.

11

THE SOUND OF WINGS

Tread but carefully
In mountain land,
Lest you sever
The unseen band,
And gather knowledge best unknown,
That drifts in sky and wind and stone.

HE was mustering alone on the back of the Forks. It was a sunny day in late fall, the sort of day when the hills are at their best — just a gentle drift of air down the gullies, the jagged peaks clear-cut against the deep blue sky, with the sun shining bright and warm on the grey of bluffs and scree, the golden tussock on the spurs flowing down to bend gracefully, in the fullness of their growth, over the softly murmuring creek.

The valley silence, filled with the undertone of the creeks and the enthusiastic chirping of the hosts of grasshoppers, was broken only by the occasional clatter of a shifting piece of shingle or the cheerful call of a kea high up in the top basins. Such a day when the expression "the music of the hills" really finds meaning.

What a change would be wrought over this country in a few short days or weeks. Soon, very soon, it would lie silent in the grip of snow and ice, a sterile, forbidding and dangerous land; it would remain so until the seasons, in their never-ending cycle, released it once more to life.

He turned away from Lake Hope, its sparkling waters cupped in virgin rock, to start his swing out through the hanging basins of the back face — even now in shadow, this was a steep and difficult block of country.

He had caught six toey wethers beside the lake; these, after being started down the valley, were now moving purposefully across the flats, as if conscious of their indiscretion in being out at this altitude at this time of year. He came on another nine and these he was content to follow quietly as they strung on ahead of him over this small bluff, along that convenient ledge and through the numerous chutes, following the well-beaten track, worn deep by generations of sheep, that led up and out to the clear country of the Front Face.

Lake Hope is at 5,860 feet: "A difficult block of country."

As he moved through the face he went out on to the points of bluffs, here and there, searching for any other small mobs left behind in the main muster — from one of these he saw, far below him, standing on the jutting top of a bluff, four sheep. He cursed the whim, or the keas, that had driven them into such a place, for this was a bad bit of country. Towering bluffs intersected by narrow tussock-covered ledges cut vertically by smooth rock chutes that plunged almost straight down, a good thousand feet, to the valley floor below — and there they were, right down in the guts of the bad stuff.

He stood and looked in disgust. Here they would be trapped by even the first light snowfall, to die of starvation or to be swept, later on, to their doom by avalanching snow. He reflected bitterly that they also stood a good chance of falling to their deaths if he tried to get them out. Should he leave them, in the hope they would make their own way out? But time was running out for the high country, the hills were unusually clear of snow for this time of year, and the first fall could come at any time. As well as this, he knew that sometimes sheep driven or strayed into such a position would remain there, apparently lacking the initiative or inclination to make the effort to get clear. Sometimes in such a case, all that was required to get them out was a bit of a stir-up; others had to be forced to move and guided out.

Well, they'd have to take their chance now, for he would not be out this way again before snow could come. First he would try a couple of "Maori huntaways" so, selecting a good-sized round rock and going to the edge of the bluff he was on, he rolled it over, on a line that would pass close to the sheep. It gathered speed instantly, leaping from the top of one bluff to another. It struck hard on the point of one bluff and leapt far out into space, high above and well out from the sheep — disappearing from sight with an ever-increasing whistling scream, followed shortly after by its crash far below.

Try another, down the chute this time. Away it went like a bullet, leaving puffs of powdered rock at every bounce. It ricocheted off a projecting bluff and sliced through the tussocks where the sheep stood, to disappear from sight below.

The only reaction from the sheep was to rush right out on to the edge of the bluff and stand in an agitated bunch. Hell, he'd have to give it a go to get them out. He started carefully down the chute, thankful for the solid anchorage of the occasional snow tussock and with his three dogs, claws bared to the solid rock, following closely. Occasionally one would slide down to bump into his legs from behind and be greeted by a string of curses that readily found their way past the heart in his mouth. The gunpowderlike smell of the pulverised rocks he had sent over drifted up to him.

As he worked his way down, he thought of that other musterer who, many years before, had also been spotting this block — he had not returned to camp that night. The others had found him next morning, just a pitiful broken heap, at the foot of one of these chutes — his dogs sitting guard. They had blamed his fall on the black ice in the chutes — that treacherous, crystal-clear ice that forms over the wet black rock with a sudden temperature change.

There was no ice today, thank God, but it was dicey enough as it was. He reached the level of the sheep and they crowded to the other end of the narrow ledge — there was obviously no way out for them that way, the only hope was to get them to scramble straight up the chute which, once they had got the idea and, given time, should manage quite well. But how to stop them turning *down* the chute when they came out? That way led to certain death as it ended abruptly in space twenty feet below.

He dropped down another five feet and tossed some small rocks behind them; they only jerked their heads and crowded further away. He was afraid to send a dog out on to the narrow, steeply

Sheep will wander even into country like this. The shepherd and his dogs must — somehow — get them out again. The Rocks block.

sloping tussock ledge, for a dog will often slide out of control on thick steep snow tussock. Apart from that, a dog could well panic them into a charge down the narrow chute, taking the dogs and himself with them into space. There was nothing left but to go in himself — it was either that or leave them there and climb away back up, with the nagging thought that he might have got them out if he had tried.

He studied the ledge closely and tried to visualise every possibility, for here was no room for error. Finally he moved the dogs to a small ledge, a bit lower down and on the opposite side of the chute. Telling the dogs to "sit" he moved up to the ledge and began, slowly and carefully, to work his way out. He would have to keep on the bottom side so the sheep could run out close into the bluff behind.

As always, in situations like this, he thrust all else from his thoughts — the sickening void of space below — even the sheep — all must be suppressed as he concentrated on finding each foothold, on ensuring each handful of snow tussock was living and solid. The slope of the ledge was such that the tussock was brushing across his face as he went — he paused and, moving the strands of snow tussock aside from above him, looked up for the sheep. There they were, four or five feet higher up and half a dozen yards further on, staring down at him through the tops of the tussocks, their nostrils expanded and twitching. They looked scared but not panicky. If only they would move out quietly. . . .

Another couple of yards and they might start out. Once more he parted the tussocks and looked up. They were still there, staring at him. Suddenly one made a lunge back along the bluff, the others plunged to follow, one, sliding off the top of a big snow tussock shot straight for him. With the flashgunlike impression of what had happened clear in his eyes, he threw himself into the slope, pressing his face into the tussocks, both hands full and tight, *tight*, on those good strong strands. With a thump the sheep hit him on the shoulders, driving him down till his hold on the tussock stopped him — it nearly tore his hill bag off as it hurtled out into space.

With his heart thumping in his ears he lay there for a few moments. Then, rising, he moved up to the bluff to sit with his back against the cold solid rock till his jumping nerves settled down. There came a rattle in the chute and looking up he could see the other three sheep picking their way up, angling from side to side, steadily, as if this was a daily routine. There came a rustle in the tussock along the bluff and his old heading dog came nosing up to him, concern quite evident in his actions. He threw an arm over the dog and leant back against the rock, his eyes closed.

His nerves gradually steadied and he was just on the point of rising when, from far down the valley, seemingly at the same height as himself and moving towards him, came the unmistakable sound of the measured beat of a pair of giant wings. Steadily it came on, the measured beats growing louder. The sound passed by, level with him and just out from the bluffs — his ears guided his eyes along its course as the sound gradually faded towards the top of valley, almost immediately to start again and move back past him, to fade out down the valley and not return.

With the hair prickling the back of his scalp he remembered he had heard the same sound, briefly, once before, in much the same circumstances.

He had seen nothing — but he wondered, as he rose to go, if that musterer of long ago had seen as well as heard.

12

ROUND THE FIRE

Laughter, fun and
Friends of yore,
Once lived behind
The now closed door.
No cheerful fire at end of day,
Now comforts friends; gone each his way.

THE old bunkhouse hugged the ground in the gully, just a wisp of smoke drifting up from the chimney.

On this sparkling clear night of heavy frost the full moon shed a light only a little less than that of day — a myriad frost crystals sparkled like scattered diamonds from the rock-hard ground. The drifts of snow remaining and the piles of it along the walls, that had slid off the roof, gleamed whitely.

The gang had just trooped back down from the cookshop after a most satisfying tea. The hut was still nicely warm but willing hands brought in and piled more logs on the big fireplace; deep and six feet wide it was ideal for the long matagouri logs, for the less axework involved in cutting that rock-hard wood was all to the good.

An ideal bunkhouse this, thirty feet by twelve with a low ceiling, all lined with T&G. The eight bunks were ranged round the walls between cupboards; there was a long lino-covered table down the centre with handy-sized forms; the big fireplace was at the other end wall where also an array of oilskins, gear and bridles hung from pegs.

Slightly over-full of food as they were, the boys did not feel up to cards at the long table so by common consent gathered round the fire for a smoke and a yarn. Conversation at such times covers a multitude of subjects. There is bound to be one inveterate yarn-spinner as well as someone who plays a good tune on an accordion or mouth organ, and one or two useful singers. This evening was no different — the yarns followed one on another in easy succession, in the ruddy glow of the fire.

"Have you heard about the time on Glencoe when the packhorse smashed the rifle and they had nothing to kill the wild bull with for dogtucker? No? Well, some of the boys had ridden out back early, so as to get in this young bull that was scarpering about with the cows on the flat.

"After several goes they managed to get him into the yards with a dozen or so cows and His Nibs was by no means pleased with the setup. He kept ramping around the yard, taking a poke with his long horns at anyone who got near the rails. As the yard was only made of beech poles and was pretty old, it wouldn't have stood up to a decent push, so they decided to keep well away and hoped he'd be still there when the packer arrived with the rifle.

"Well he was, only just, still there when the packer got in, but this joker had slipped the rifle through the pack straps on the wrong side of the load, and the horse had caught it on a rock coming up the gorge and damaged the bolt. No one felt like pulling the trigger, as it was anyone's guess as to which would travel the fastest, the bullet or the bolt.

"So here's this bull still circling the yard, looking for the gap he might have missed in the thousand or so circuits he had already made. The boys grouped outside the hut eyeing him and trying to work out how to do him in. There was no rope or even wire that they could lassoo him with so it was some problem. If they let him go, it would mean someone would have to ride out the thirty-odd miles later to drop him — that's if they could find him, after he'd been disturbed like this. They would probably never get as good a chance at him again.

"Some pretty bright suggestions were put forward — like someone (not the bloke who was suggesting it, of course — he was always pretty stiff after the ride or had a weak ankle or something) could creep up on His Nibs and stab him in the heart with a knife tied on the end of a stick — or two could grab him by the tail while the others cut his throat — and so on.

"Time was slipping by and they were getting nowhere fast. George, the head shepherd, was getting a bit hot under the collar, what with all the bright suggestions and general state of affairs. Then 'Right,' he says, 'I'll pith the bastard,' and explained what he had schemed out.

"The yard was built in among the beech trees and quite a few big boughs hung out over the yard, low down. His idea was that he would try and get out, unseen if possible, on to a good low one, then drop on the bull's back and pith him. The boys took a pretty dim view of this, but there was nothing much they could do to stop George, short of tying him up. He was the boss, and tying him up might be a bigger problem than tackling the bull — so they got serious and planned things right out: one of them would wander, casual-like, down through the trees at the other end of the yard, gradually drawing closer and trying to hold the bull's attention but without spooking him. Meanwhile, George would slip down

Polled Hereford bulls on the prowl.

through the trees, then up one and out on a bough.

"The others armed themselves with good solid waddies, the axe and a bit of standard, to ward the bull off George if things went wrong and he got caught in the yard. Those cows weren't much encouragement either — they were all big, old-style Herefords with a fine spread of horn and not what you would call hand-reared, they hardly saw a human all year and very rarely on foot. No one was ever keen to go near them except on a horse, and they were sure to take a pretty dim view of George dropping out of the sky, smack in among them.

"Well, everyone was pretty uptight, fully expecting to have to rush in and protect George and not exactly looking forward to it. What with one thing and another, they reckoned the sooner they got it over and done with the better. Away strolled the first bloke through the trees. As soon as the bull spotted him, he stopped his circling long enough to toss a few barrowloads of dust over his back and give a few highpowered snorts, then back to his circling, only halfhearted now, keeping a watchful eye on the bloke strolling closer through the trees.

"While the bull was busy showing off his style with the dust and

A note scribbled on the back of this photograph reads: "S. Stevenson and G. Metzger with Rita (behaving normally)". The New Zealand shepherd's life has its Wild West moments, but it has not — as yet — been Hollywoodised.

more or less blinded with it, George nipped up the tree with his knife in his teeth and crept out on the limb — the bull circled once, twice — then George jumped.

"*Wop* he went, right astride the bull's shoulders and, almost before he touched down, *WOP* went the knife, right up to the hilt behind the bull's head. The bull dropped in his tracks in the dust and George rolled clear right smartly — the cows took one look at George and didn't even notice the other side of the yards was there as they went straight through, throwing a scare into the bloke outside.

"George stood there, as cool as you like, brushing the dust off his pants and calling to the boys, standing outside with their mouths open, to hurry up and come and cut the bull's throat."

"Do you mind the time, that dipping muster two years ago, how hot it was, the day we did the Back of the Neutrals and James Peak block?

"How we'd had to pull back to the Two Mile hut because of fog, the day before? The loss of that day left us with just enough tucker for the night at the upper Staircase and a round or two of bread to last us for the Big Spur next day, joining the packer at the lower Staircase that night.

"Well, when we reached the hut the sun was still blazing down and not a breath of wind. The hut was like an oven, so we took the tucker outside into the shady side and cut up the last of the meat and bread there, while Steve went to bring the butter back from the creek — it had been so hot, most of it had melted and run round his bag and he had taken the bit he had saved over to the creek to harden up. He seemed to be a way a long time, then he sang out he couldn't find it and that it must have got washed down the creek.

"Not thinking, we all left our plates and went to help him look for it — not a sign of it, till someone spotted a bit of paper. The dogs had found it and eaten it. With everyone blaming each other's dog we went back to the hut — the flamin' dogs had been there while we were gone. Not a thing left, the plates licked clean and we hadn't had the chance to have a bite — our bellies were sticking to our backbones by the time we got in the next night.

"It was the fastest that block was ever mustered — that was the day that new chap had the job of turning the mob down into the crossing in the bush. He was standing waiting on top of the spur, at the head of that practically vertical chute you slide down on into the bush.

"The lead went pounding by and he took a leap into the chute — we saw his stick go twirling up into the air and a cloud of dust shot down the chute. Cripes, we reckoned, he must be in a hurry — but no, he staggered out the bottom and sat down on a tussock. When we drew level with him he was a proper sight, you'd have thought someone had run over him with a set of harrows.

"What had happened was, he took a mighty leap into the chute all right, but one of his heel plates was loose and caught in the cuff of his other pants leg and tied his feet together — he had gone down head over heels. We told him he must have set a record time for going down the chute, but he didn't seem to be over-delighted.

"This chap must have been born lucky, if you can use the term that way. That same muster, the day we did the First Day in The Rocks, he put the wind up us — all hands had got clear of the bluffs on to the tussock below and were waiting for him to show up. We finally spotted him, high up in the top of the bluffs. He had mistaken the chute to come down and had got a mob of about thirty bluffed on a steeply sloping ledge covered with heavy snow tussock.

"We could see him surveying the situation and trying to work out the best way of getting them out. Well, none of us envied him the job, he was right up near the top of the worst possible part of the face, which was an almost vertical series of bluffs for at least two thousand feet. The bluff the sheep were stuck on was a beauty, six hundred feet of nothing but clear mountain air for a starter, with only a few ledges jutting out from the tops of the other bluffs right to where we stood.

"Six hundred feet of pure mountain air for a starter." First day in the Rocks.

"This bloke had done a fair bit of mustering but this was his first time on really rough country. The head shepherd was looking a bit anxious and tried to roar out to him to leave the sheep, but the wind was too strong and he couldn't make himself heard. All we could do was to stand there, gazing up and getting a kink in our necks while we waited developments. We all knew what it would be like up there. The snow tussock would be slippery as glass, the void below would smartly tie a knot in your bowels, and the wind would be whistling straight up the legs of your trousers — the sort of place where you didn't make one move without a good handful of old-man tussock.

"Well, this bloke was game all right. We could see him trying to get his dogs to sit down back out of the way, then he started edging round through the tussocks below the sheep. The head shepherd was cursing quietly and there was definitely no light conversation going on among us. We could see him edging round, occasionally slipping in the tussock, until he got right below the mob when, suddenly, his whole pack of dogs tore round and down on to the ledge.

"The mob bunched, hestiated, then took off straight down, across the ledge, over and around this bloke, and he was lost to sight — the sheep just ran straight out into thin air, to come floating down towards us in that deceptively gentle sort of way that is only dispelled when they hit the ledges with a sickening *wuff,* to go bouncing out into space once more.

On a day like this the lakeside visitor will curse the weather — the shepherd does likewise, but gets on with the job. The Devil's Staircase, with a big drop below.

"We thought he was a goner for sure, for with all those falling bodies, floating and bouncing down, it wasn't possible to say whether he had gone over or not. As what remained of some of the sheep bounced to rest in the creek beside us, there was still no sign of him — only his dogs lined up along the back of the ledge.

"We were feeling pretty sick in the stomach and not looking forward to having to search for the pieces, when the flattened tussock on the ledge parted and, thank God, we saw him climb slowly up to his dogs. He said afterwards that it was only the long tussock pressed down over the top of him that had let the mob slide over him, without tearing him loose."

If there happened to be any new hand or stranger present, the yarns seldom ceased without someone bringing in the story of a head shepherd in the early 1900s and it would go somewhat like this:

"You'd be wondering why anyone would be silly enough to leave that gate unfastened between those two paddocks of newly drafted sheep which we had to redraft this afternoon? Well, that gate was fastened all right, but Old Tom must have been on his rounds again. Who's Old Tom? Well, I'll tell you.

"Tom was head shepherd here years ago and he was a first-class hand. Everything was done properly, no slipshod work allowed, and he was a good man on the hill.

"Well, it had been a pretty good season on the station with no more than the usual number of hitches, but there'd been an early snow before the fall muster and it carried on with more skiffs, gradually building up until there was a heavy coat all over the high country. They got the fall muster done all right and waited a bit for it to clear to do the spot muster, but instead the snow kept coming and they never did get it done.

"The result was that they were down a bit in the tallies. Nothing serious, but it must have been worrying Tom more than anyone guessed. All hands were sitting yarning one night round the fire here, just as we are, when Tom got up quietly and went out — no one gave it a second thought — but time passed and he didn't come back. Someone reckoned he might have gone to look at the horses and found something wrong, so a couple of them went out to have a look, but there was no sign of anything wrong — and no sign of Tom, either. They checked whether he might have taken his dogs to do something, but they were all tied up. They reckoned he might have taken it into his head to walk up to the homestead to see the boss, so they let it go at that.

"He hadn't returned to the hut when they awoke in the morning and when the boss came down he said he hadn't seen him either — so they started a full-scale search. If he'd slipped on some ice and hurt himself he'd have had a pretty tough night of it in the frost.

"Well, they hunted high and low but they never found a sign of him and all their enquiries at Queenstown, the Arrow, and Kingston and Cromwell showed he hadn't lit out.

"He'd just got up, walked out and disappeared. Left all his gear, money, dogs, horses, everything. . . .

"Time passed and everyone came to accept he had walked out and committed suicide in such a way that he could never be found.

"We all reckon his ghost comes back again occasionally, and if he finds something done that he doesn't agree with he gets it done again by opening a gate or two — a real stickler for doing things properly was Old Tom.

"You wait. If you're here long enough you'll get checked over by Tom."

A replete silence settled on the group, as they lounged on the forms in the ruddy glow of the fire — each following his own thoughts — or none — in the tumbled gorges of the glowing coals. Sparkling lines climbed and spread across the back of the sooty black fireplace, for all the world like tussock fires on the hill at night.

Someone leant forward and pushed home a twisted, black, six-inch thick butt of matagouri that was threatening to topple out, and, as a flock of fine sparks rose swiftly to disappear up the maw of the chimney, said, "Talking about walking . . . ."

"Who the hell's talking about walking," said another, "that's about the last thing I want to talk or even think about."

"Well, it's more about a woman, really, and . . . ."

"Time you had a bit of exercise," put in the head shepherd, "you can nip up to Brown's Basin in the morning and see if the keas have been working."

"'Struth," in desperation, "you blokes could have been born under a matagouri, for all the manners you've got. What I wanted to ask was, does anyone know anything about that heelplate of a woman's boot that is nailed to the Wye Creek hut door?"

They looked expectantly at the head shepherd — "The boss told me the story behind that — hang on till I fill my pipe and I'll tell you the yarn."

He carved a generous fill from his plug of Golden Eagle and ground it carefully between palm and heel of hand — the rather evil looking short bent pipe was lovingly packed and, leaning forward, he plucked a glowing coal from the fire with his bare fingers and applied it to his pipe. A few deliberate puffs and the pipe was alight and drawing well and the coal was tossed casually back into the fire.

Concealing his awareness of the not quite wide-eyed regard of the youngest member present, (the reaction to this performance is always worth the effort) and, suppressing a chuckle, the head shepherd settled back and said,

"It was a long time ago, apparently about the time the first gold boom in the district was petering out, when this bloke and his wife and four-year-old daughter came up from the south to try his luck with the gold.

"They started prospecting from the south end of the lake, working their way up round this east side. They evidently got a few colours at Clippings and the Staircase, but only enough to convince him there was nothing payable there.

"They moved on up the side of the lake, a day here and a night there; him carrying their tucker, blankets, a tent fly and his pick, shovel and gold dish, which served as pot and camp oven as well. The woman carried the little girl.

"Eventually they got up as far as Wye Creek and set up a more permanent camp, just above where the hut is now — you can still see the rough stone fireplace. They stayed there about three weeks while he prospected around, but he didn't get anything more than a colour. It's a funny thing that while there has been plenty of gold got on the north end of the Remarkables and down the river, only the odd colour has been found down the west side.

"Anyhow, one of the station shepherds happened to be passing

"That pass through Ben's Basin is bloody murder . . ." Looking back towards the pass during the spot muster. Brown's Basin lies just below the horizon to the right of the Cones.

through and met this chap. He told him they had been living mostly off the land, eels, cabbage tree heart and fern roots, and that their bit supply of flour, tea etc. had just about run out. If he didn't strike anything within the next few days he was going to prospect right up the Wye, out over the tops and down to Cromwell where he reckoned there would be a chance of a job on one of the big claims.

"The next time someone was down that way they went by the camp — there was only the ridgepole and the bough sides standing. The fly was neatly rolled up and lying inside, with the pick leaning up against it — stuck under one of the ropes of the fly there was a note saying, 'Travelling light. Gone out over the top to try our luck'.

"No one ever learned whether they made it or not — but I think they would. They must have been as tough as nails, but, Lord, it must have been a long hard haul. You know what that country is like.

"Years afterwards, one of the boys came across that heelplate right up near the Bridal Veil Falls — he brought it down and nailed it to the door as a tribute and a sort of memorial to the woman."

No one spoke for a while, as each in his mind's eye pictured the struggle the couple would have to rise the first steep pinch; then would come the long valley with another sharp climb at the Bridal Veil Falls and at its top the wide expanse of Brown's Basin with its bench after bench of pale grey rock and the tarns, so clear that unless there came a glimmer of reflected light you were not aware they were there. Their waters would have no taste and be as cold as ice, their surface only occasionally rippled by that thin cold drift of air that is not entirely absent even on the calmest days — the small child would hear the sharp drawn breaths of her parents after the steep climb up the black rocks beside the falls, their lungs straining for oxygen in the thin air. She would sense the brooding silence of the high places and tighten her arms around her mother's neck; but there was bound to be a wee pipit to bring a smile to her face — sitting unafraid on a rock he would be, cheerfully bobbing his tail, and giving her a bright knowing glance as they passed by.

They would be glad to reach the far rim of the basin and see the plains of Central Otago stretching warm and brown into the east. Far to the north, Mount Cook with its attendant peaks would be floating above the blue haze of distance — but they would hasten on, either down the long ridge to where the Nevis River joins the Kawarau, or take the long sidle through basin after basin to cross in the Nevis Valley and drop down direct to Cromwell.

Which route they took, no one knows.

"Who was this Brown the basin is named after?" someone asked, breaking the silence.

"He was a summer boundary-keeper," one of the old hands replied, "used to camp in that cave near Lake Alta below the Double Cone. His boundary was Doolan's Saddle through to Lake Hope, where he linked up with the other summer boundary-keeper who camped in the Two Mile Basin."

"I asked a chap who worked here years ago about him. He said he was a solitary old cuss. Used to take off for the tops as soon as the saddles showed any signs of the snow clearing. Came back down to the flats for a while during shearing and dipping and that would be the last they would see of him until the fall; apart from the occasional quick visit for rations and tobacco.

"He evidently had wonderful heading dogs, but his pack was as wild as wolves — would just about pull another man down if he came on them suddenly."

"If an old chap like that can live up there the country can't be as tough as you blokes make out," said the teamster who had joined the group. He swung a billy of water on to a hook above the fire, ready for the evening cup of Camp coffee, and took a seat on the end of a form.

The head shepherd looked at him thoughtfully, then gave Bob a sly glance, "Well, I suppose we are bound to exaggerate a bit, aren't we Bob."

Bob gave a noncommittal grunt and gazed moodily into the fire.

"Tell him about last shearing muster."

"Hell," said Bob, fidgetting, "that was nothing."

"Go on, unless you would like me to tell it . . . ."

"Aw, I suppose I'll have to," said Bob hurriedly.

"It was like this. We were mustering the Lumber Box and I was top on the Back — you know it was the first time I had been on this country. Well, the snow had gone back pretty fast that season and the wethers were fairly well out, there were even some below the Cones and we had had to take in most of the full day's muster.

"Most of you know what that beat is like. It's bad enough all the way but that pass through the back of Ben's basin is bloody murder, the first time through. It's a dark face all the way, never gets the sun, and the rock of the pass is smooth and greasy as black glass. Slopes the wrong way too, tilting down so you can barely stand on it, even in summer. It must be two hundred yards long, the actual pass. Starts off all right, about ten feet wide with a bit of loose rock on top. Just as well, because it's about a four hundred foot drop there.

"You sail on in thinking it's a bit of cake, but the pass narrows steadily and the rock slopes more. As you near the end you are sidling along with your belly pressed against the bluff, your hands stretched out slipping over that cold wet rock seeking the grips that aren't there. The only thing that keeps you going is the fact that you're just about the end — and it's an awful long walk round if you go back. The drop is only about fifty feet now, but, man those rocks below look awful jagged.

"Finally I got through and, sweating a bit, (it's dead right what they say about breaking out into a cold sweat; had never believed it before) stopped and had a good look back at that pass. It looked quite easy, but now I knew how easily that good huntaway called Ben had been lost there and quite a few sheep from time to time.

"Way down at the bottom of the basin the boss barked up and moved on. He had been watching me through. Compared to the beat that lay ahead of him, this pass was chicken feed.

"Well, first time along that top beat is no cake walk either, and I was far from pleased when a puff or two of fog formed, then set in solid, a real pea-souper. The few sheep I had in front of me clattered off into the murk. From way down in the chutes came the sound of the boss's dogs. It was impossible to see more than a dozen feet.

"It looked as if we were in for a bit of a wait, so I sidled round to where a big rock jutted out from the hill, made my way out to the point and plonked down with my legs hanging over the edge. I took off the hill-bag and started in to my lunch.

"I was reading the lunch paper when I noticed a bit of a breeze had set the fog drifting, and happened to glance down over the top of the paper lying across my knees — hell, it still makes me sick in the stomach when I think about it and if I'm tired I see it again in nightmares — the fog had drifted apart forming a clear shaft, and I was looking straight down, with nothing but a few wisps of drifting fog between me and the valley floor, well over a thousand feet below.

"I tell you, my hair stood up like the hackles on a dog's back and I froze solid. Anyone could have pushed me over and I couldn't have moved a finger.

"The fog closed in again and I relaxed a bit, bundled my lunch up and carefully pushed it out of the way behind me. I had just started to move very carefully to get back off that bluff, when the damn fog cleared again — my guts screwed up into a knot and the sweat poured out of me. I daren't look down and I was frightened to close my eyes — to look at the drifting fog around about made me feel dizzy. One of the bloody dogs came and grabbed my lunch and I couldn't do a thing, not even roar at him; I could hear the boss barking-up way down below the soles of my boots, but I wasn't game to bark up in reply in case one of the dogs jumped up on me.

"I leaned back on my hands and carefully edged well back until it was safe to stand up, then went and sat down with my back to a rock and had a fit of the Joe Blakes.

"I could hear the boss barking-up occasionally and I answered him — the sound was getting closer and I realised he was climbing up to join me. Eventually he pinpointed me and came climbing up out of the fog.

"When I told him what had happened he thought it was a great joke. Told me I had picked one of the worst overhangs on the whole face to have my lunch on.

"He had decided to abandon the day as we had lost too much time with the fog, and would only get benighted if we carried on.

"He was going on up, and out over the Rough Creek saddle to collect the boys off the Front Face and go back to the Wye camp that way, so I tagged along, rather than tackle new country in fog. It's hard enough to keep on the right line even if you know the country, and I was still a bit hesitant about stepping off any good sized rock, just in case it didn't have a bottom to it.

"No, the country isn't tough, it's just steep as a hen's face. Who's gonna make the coffee?"

They had finished their coffee and as the new hand began washing up the mugs (it being his turn), they reluctantly withdrew from the fire and made for bed.

The big central kerosene lamp was blown out and candles flickered to life at the individual bunks — the head shepherd lit his small bedside lamp and, all unconsciously, began the nightly routine that had continued as long as the old hands had known him and which was a source of mild amusement to the others.

First the diary would be dragged out and the events of the day entered up, then the left boot would be unlaced (always the left one first), removed, upended, and the heavily nailed sole inspected for any damage. It would then be dropped with a crash on the floor, shortly to be followed by the other after a like performance.

From long habit the same procedure was followed irrespective of the type of footwear in use at the time, be it good shoes or even slippers — the boys idly conjectured as to what would happen should he ever start for bed bare footed.

As the diary was slammed shut the new hand asked, "What's this post office Bob was talking about?"

"It's the first spelling place for those on the climb out to muster either the Back of the Forks or the Back of the Two Mile — it's just a big overhanging rock near where you hit the creek after going through Hawke's pass. It is pretty good shelter and mighty handy if you have to wait to see what the day is going to do." — pause, as a contemplative look is taken at the shining nails of the left boot — then, CRASH, and "Many's the snowy wait has been put in there. Everyone stops there and it is the recognised thing to put your names on a bit of paper and stick it in the big crack at the bag of the rock along with the hundreds of others that are there.

"If the weather is doubtful, quite a bit of time may be spent waiting there and someone is sure to record a bit of news or a poem and leave it as well. It used to be quite interesting to take out a handful and read it over — unfortunately some dammed kea or 'possum must have got in and scratched most of the 'mail' out on to the floor of the cave, and when we burnt the Two Mile block the last time, the fire went right through and took the lot" — CRASH, went the second boot.

"It was a real shame it got burnt. There were names and bits of news and records of climbs and blokes getting lost in fog and heavy snowfalls, right back to 1902 — even the record of Ford's death in the chutes at the back of the Forks. We always meant to bring it all in sometime for safe keeping — but somehow it didn't seem right to shift it — and now it's gone."

13

KEA, KEA-AH

They live a life
By Nature's plan,
And care not for
The designs of Man.
Each a link in Nature's chain.
Is Man a link and over-vain?

THOUGH at first sight the high-country lands appear to support little or no birdlife, closer acquaintance reveals that this is by no means correct. Thinly scattered it may be, but it makes more than adequate amends for this in its variety.

Under their winter coating of snow the high lands give little indication of the rich and abundant store of food available to the different species of birds during the summer months.

Each one wonderfully adapted to its chosen environment and way of life they all, excepting the kea, tend to follow a semi-migratory pattern in altitude — even the kea, that astounding character, condescends to widen his foraging range in winter.

Seldom is it that one leaves the southern alpine country without seeing, or at least hearing, the kea. His ringing cheerful call of *kea, keee-ah* echoing among the crags is as much a part of the unique New Zealand scene as the song of the bellbird in our lowland bush.

He has chosen some of the roughest and climatically savage country in the world as his natural habitat, preferring to live above the bushline in the Southern Alps.

The entertainer and casual companion of the mountaineer, the bane of the high-country runholder's life, an aerial acrobat, clown or sadistic devil — cheeky, inquisitive, destructive — he is all these and more. If he had ears they would be bright red, like his underwing, with the names he has been called down the years. The stories about his exploits are legion. There must be no hardier bird anywhere in the world and certainly not many more colourful — in either plumage or character.

He is a member of the parrot family, named *Nestor notabilis,* and his size is about that of a large pigeon. A powerful and most useful hooked beak, brown to black in colour, tending to yellow in the younger birds. A cere and eyelids of yellow, set off a pair of intelligent, sparkling brown eyes. Main plumage is of various shades of iridescent green, feathers edged in black. Underwing and

"Cheeky, inquisitive, destructive . . . They tread the most powerful winds of the mountains as an athlete would his trampoline." Some keas are killers of sheep, too.

body coverts are bright scarlet. A pair of powerful feet, with two claws in front and two behind — ideal for climbing and hanging on to rock faces, but a distinct disadvantage when walking, when he resembles nothing less than someone trying to walk on bare ground equipped with a pair of short skis. The toed-in, high-stepping walk becomes an asset in his clowning acts.

His mode of flight on shortish, broad, powerful wings is more businesslike than graceful and his complete mastery of his element within his chosen environment must rank with the albatross' in his.

When a flock of keas takes off from some snow-covered hillside, one would almost think flashes of flame were being struck from the rocks, so brilliant and surprising are their under coverts; but to really appreciate their flying skill one must see them among the bluffs in a high wind — they are never the playthings of the elements, rather are the elements their plaything.

To see keas hanging casually in space, on practically closed wings, in the worst winds and weather nature can attain, swinging in obvious enjoyment, up, down and across the face of some huge bluff, then to go with a joyful scream, shooting like bullets on the wings of the wind, to land with accuracy and aplomb on some tiny ledge, is a sight one will never forget. They tread the most powerful winds of the mountains as an athlete would his trampoline — for his own use and entertainment.

Their natural foods are anything edible, but mostly roots and insects. That they will and do kill sheep is fact, and their wanton slaughter has brought ruin to many a high-country run. That this wanton slaughter is usually the work of a few individual killer birds is no consolation to the runholder.

The real killers usually work alone or in pairs. They are relatively easy to detect; usually older adults, they keep, or are kept, apart from their family flock. They are, almost without exception, silent and watchful, conveying the impression of evil as if obsessed with the desire to kill, like a rabid dog.

The kea shares the mastery of the high-country skies with another, that he treats with due respect — rejoicing in the name of *Falco novaeseelandiae,* commonly called the sparrowhawk in the south, bush-hawk elsewhere.

Here too, the New Zealand falcon's wild call is very appropriate to the surroundings. Consisting of a high-pitched, staccato series of *se--se-se-se-se-se-se,* he throws out a challenge to all and sundry as he wings swiftly among the crags.

Undisputed fighter ace of our skies, a fair comparison with the kea would be as between the sleekest jet fighter one could name and a specially designed aerobatic sports aircraft.

The writer is not aware of any instance where the falcon has attacked a kea, but the respect in which he is held by them is quite evident should he pass over a family group of keas — their jobs in hand are temporarily abandoned and, with much cocking of eyes skyward and various lewd comments, he is kept under close observation until they are satisfied he is "just passing through".

Approximating half the size of the harrier hawk, that cowardly soarer of the foothills and flats, the falcon is a beautiful bird. Designed and perfected in the forge of time, every line of him is smoothed for least resistance and drag. His upper plumage is warm to dark brown with a smooth helmet of darker tone, wing and tail feathers barred with pale rufous, underbody buffy white, pale yellow legs and feet and those terrible weapons, the talons, black. A bluish-black beak short and sharp, a yellow cere and a pair of very alert, yellow-irised eyes. Wings are narrow and tapered as befits a high speed package.

His normal flight is usually in fine pitch, a steady series of short sharp beats. When full power is required he passes to coarse pitch with powerful enlarged wingbeats that increase his speed threefold.

His original hunting range was all of New Zealand, but he has now been driven back into the bushclad foothills and the high alps. Sad to relate it would appear that he is on the way to becoming if not an extinct, then a very rare species, unlike the kea who, safe in his mountain home, continues to flourish and increase.

New Zealand falcon: "Undisputed fighter ace of our skies."

This decrease in falcon numbers can be accounted for in two ways: first, the widespread use of chemicals such as DDT which, concentrated in his prey, reduce fertility; and, secondly, senseless and indiscriminate shooting — for this bird knows no fear and recognises no master. Discretion yes, but fear, never. He will make a low pass at a human just for the hell of it and will attack fiercely, even strike a human who comes too close to his nest. He will glide smoothly in to land on rock or branch within feet of a party of trampers or shooters, to sit and survey them from calm and fearless eyes; unfortunately humans do not abide by the laws he follows — and might be the better if they did.

The falcon never kills for pleasure and the majority of his challenges are to even combat in the air — other birds appear to know when he is on the hunt, at times evincing no more than sharp attention should he land near them. When he is on the hunt there is a mad rush for cover.

He is very fast in level flight, but it is his speed in the stoop that is so amazing — the English falcon has been accurately clocked at something over four hundred miles an hour in the stoop. Our bird, being smaller and lighter, would probably not attain more than half that speed.

On sight of prey there comes the flick roll into the stoop, the second lid comes over the eyes to shield them from air pressure, a few full-power rapid wingbeats, depending on the speed required, then — straight down on the line of attack, the wings almost closed and, travelling almost too fast for the human eye to follow, this avian projectile is in full stoop. The talons come out in the last split second — a puff of feathers and the prey tumbles, lifeless towards the ground, to be struck again and carried off before reaching it. Size alone is of no protection to the prey, for the power behind the strike centred on those lancelike talons would carry the shocking-power of a bullet.

It is difficult to translate the kinetic energy of the strike into easily appreciated terms, but an assessment of the power of an average strike can be expressed and appreciated as exerting a force of over four hundred tons to the square inch.

The falcon's way of life has been termed cruel but, as a little reflection with show, it is no more so than that of any other predator — he is an essential part in maintaining Nature's balance, even to keeping the little mountain pipit population within bounds.

In practically every area of New Zealand the little pipit can be found. Similar to and related to the common lark they especially frequent the warmer tussock slopes of high country, up to and above the 6000-foot level, and occasionally one will put in an appearance in heavily snow-covered country.

Any pause on a sunny face is almost certain to attract the presence of these delightful birds. Mildly curious and trusting they will approach within a matter of yards, in the curious butterfly-like flight they use at ground level. They are surprisingly fast and agile fliers when making a passage and adopt a volplane type of flight of alternate strong wingbeats and closed wings.

They prefer to remain close to the ground and spend most of their time running through the tussock and rock in short dashes after insects.

Mainly silent, they do at times maintain a widely spaced series of softly modulated *pi-pit pi-pits* — their song consists of a long, barely-heard trill, usually performed from some favourite rock. Meanwhile their tail, slightly longer than a lark's, maintains a spaced, metronome-like, involuntary flick.

They follow the snowline down and back, spreading out over the lower ground in winter.

In the upper limits of the scrub, especially in the valleys, can be found the much smaller birds — the grey warbler, the waxeye and, at the top of the bushline, the rifleman. Lower down, the tomtit and fantail.

In the higher valleys up to the 5000-foot level, the watchdogs of the high country frequent the creeks and ponds. The deerstalker who can stalk his game in a valley free from the presence of the paradise duck is fortunate indeed — with eyes like a hawk's they very rarely fail to advertise the fact that there are "strangers present".

The paradise duck chooses a well-concealed nesting site in the high country, usually in a narrow cave well back under a rock. Here the nest is formed from a considerable amount of down plucked by the female from her own breast — the eggs are always carefully covered with this when she leaves the nest.

During the hatching period the male frequents one particular site, usually beside some small body of water, often miles from the nesting site. Here he potters about awaiting the arrival of the female, which is as regular as clockwork. She calls on nearing the meeting place and he takes wing to escort her in. After feeding and bathing for about an hour, she returns, escorted back by the drake, who ensures she enters and becomes established on the nest, undisturbed.

He remains in the vicinity only a short time, then returns to the meeting place.

When hatched the ducklings are brought to the previous meeting place, which has been selected as providing just the right depth of water in which the ducklings can dive to escape the attentions of the roving harrier hawk.

Considering the location and distance between some of the nesting sites and the meeting places, there is little doubt that the most common method of ensuring the ducklings' safe arrival is a swift flight on the back of the female, safely snuggled in under the feathers. There is little doubt that these ducks mate for life.

Some of the small lakes at altitudes approximating 5000 feet attract large numbers of black-backed gulls during the nesting season, evidently lured to the location by the hosts of crickets. These too can raise the blood pressure of a stalker as they follow him up a valley, announcing his coming with continuous loud cries.

It is not possible within the scope of these brief sketches to do more than make passing mention of the various species and to touch on their fascinating ways of life — merely to indicate that our alpine skies are far from empty.

Any discussion of birds naturally requires some consideration of their feeding habits — and here we are led to enter the strange and wonderful world of entomology.

It is a source of wonder to those who have once viewed the abundant and diverse insect life abounding in the high country in summer, that it could ever return after the long siege laid by the winter snows — but return it does.

The impact of our development of the lower country on the indigenous insect life must have been tremendous. The clearing, cultivation, grazing and widespread use of insecticides completely changed their environment. This could have no other result than the complete eradication of some species and their replacement with types better suited to cope with the new environment. But the high country has been spared such sweeping changes, as have the great stretches of native bushland.

For countless centuries the tide of life has rolled on unchanged by other than climatic variation; the insect life that has adapted to the harsh environment of the high country remains unchanged.

Many of the high sunny faces support scattered colonies of mountain wetas — it is not unusual for a musterer to confide to a beginner that his huntaway became a bit tired today, so he slipped a dog collar over a big weta and led him along — his barking saved the dogs no end. *Pharmacus montana,* or the "Mount Cook Flea", is a most fearsome-looking animal (his size almost entitles him to that title). This is a slight exaggeration, but they do grow to over an inch long and three-quarters through — they often come to the mouths of their burrows and "bark" if one happens to disturb their slumbers. The sound is of course produced by their rubbing the spines of their hind legs against sounding-plates on their bodies — to the other insects they must appear to be dinosaurs.

The grasshopper species are many and varied. On sunny days their constant song forms a background to the other mountain sounds. Almost invariably present too, are the little black woolly-bear caterpillars of the moth and butterfly population, always busily going some place else. Numerous small butterflies abound, with the occasional Red Admiral being carried through by the wind.

To the unobservant the mountain lands must appear empty; yet they are abounding with life.

14

THERE ARE CATTLE IN THE VALLEYS

Now saddle up
And ride with me,
We'll comb the spurs
And valleys free
To mob them up and drift them down
And draft the steers to drove to town.

THE first cattle to reach the Wakatipu were those of a Mr Few. These remained only a short time before being withdrawn. The second mob to arrive were the milch cows brought in by W. G. Rees in 1860 — the price paid for their introduction was to be high, for it cost a man his life.

John Gilbert, a relative of Rees, and J. H. Stringer, Few's stockman, journeyed from the steading at Queenstown to pick up the cows from near the Lindis. As they were crossing the Clutha on their return journey, Gilbert was swept from his horse and drowned. Recognised as one of the great rivers of the world, with an outflow of 2,000,000 cubic feet a minute, it is little wonder that the turbulent, icy Clutha posed a major hurdle to the early settlement of this inland area.

The great valleys of the major rivers draining the catchment have for many years supported large cattle herds — today, the number carried within the Wakatipu basin exceeds 25,000.

Generally speaking, cattle numbers were comparatively static until the late 1950s, when aerial oversowing and topdressing created a requirement for additional numbers. This applied more especially to the high-country runs in the district and this station was no exception.

Judging by the past records, at no time until then did cattle play a significant part in its economy. This is not the case today, 1,500 head being carried inclusive of a large Polled Hereford stud. On the gentler southern slopes cattle will graze beyond the 4,000-foot level and there is no question as to the improvement in pasture quality and coverage derived from their introduction.

Before the road was opened from Kingston, all stock being sent south for sale from the area had the choice of two routes: overland by the pack track along the side of the lake, or as deck cargo on the lake steamers, the *Earnslaw* being able to carry 1,500 sheep or about 30 head of cattle, the cattle being unable to use other than the foredeck. This resulted in any fair-sized mobs of cattle having to risk the pack track, lined with poisonous tutu plants and including the dangers of the Devil's Staircase, a face of sheer rock, through which a narrow track had been laboriously cut.

Stockbuyers from the south, wishing to make as few trips as possible, had a flair for collecting a miscellaneous mob, good steers, old cows, brokendown bulls and yearlings all together. This did not make for easily driven mobs and assistance was usually sought to negotiate the worst section, the seventeen miles of station land.

This was readily given, for the Skipper had no wish to have some nondescript bull turn up among his cows or to be burdened with assisting with the cutting-out of boxed station cattle.

One buyer passing through had, to his regret, purchased a vicious Jersey bull. He managed to get him as far as the station holding paddock when the bull, obviously seeing no future in a long trek over rough country with nothing but the freezing-works at the end, decided to go berserk. Anything that looked remotely like a human, whether on horseback or not, was immediately charged, intervening fences being of no concern. After a couple of fences went west, the Skipper laid it on the line; either the bull must be destroyed or the buyer must pay for any fences damaged.

The buyer's decision was immediate, for he and his horse had already been bowled over once, and the young shepherd was sent to get the only rifle handy. This was an early model .303 with a short sawn-off barrel and no front sight. Normally carried out to camps on the pack loads for killing dog tucker, it was very much the worse for wear, being far from accurate at distances of over six feet.

Now this bull was no ordinary specimen. At one time he had been graced with a substantial nose-ring, but at some time in his eventful career this had become hooked over some immovable object and, being the strong if not silent type, the bull had simply moved on, leaving the whole of his nose and upper lip with the ring. Now frothing at every bolt-hole, so to speak, he was not a sight that engendered feelings of confidence.

The young shepherd crept up the side of the gully and, lying amongst the tussocks at the lip, was within twelve feet of the fence which the bull was even now considering as just another minor impediment to conquests further afield.

As the bull, now ranging along the fence, drew level, a nicely angled shot behind the ear was selected and the trigger squeezed. There was a loud click — the damn thing had misfired. The bull heard the click, wheeled and, spotting the hat sticking out of the tussock, charged. The twang of breaking wires would have brought joy to any rock guitarist's heart, but it also ensured that a fresh

Polled Herefords. The Remarkables Station operates a large stud of these cattle.

cartridge was slammed into the chamber at record speed. A hasty instinctive aim for the brow and — *bamm!* — the job was done.

The memory of that horrible head, now staring at him through the tussocks not six feet away, ensured that that young shepherd had a soft spot for a good rifle in the years to come.

When Dickson Jardine took over Glencoe Station in 1921 cattle prices had been at low ebb for many years. This meant that on a number of high-country properties where cattle were carried, economics had dictated that they be left very much to their own devices, to run wild and multiply in the wild gorges of the back country.

This was the situation on Glencoe, where a large number of cattle were running wild in the difficult North Branch area.

The previous owner, some years before, had mustered in a good cut for inspection by a buyer. As usual on such occasions, bargaining took place to reach a price acceptable to both parties, and they eventually reached a stage with an offer of £18 standing against an asking price of £18 10s. There they stuck. When further negotiations failed to bring about agreement the buyer rode off in a huff. The owner was to regret not closing the deal, for cattle were shortly to slump to such an extent that they were not handled again thereafter.

All these cattle living the easy life on the property in no way met with the Skipper's approval, and a cleanup was organised in short order. It was a major operation rounding them up out of the rough, scrub-covered gullies. Older scrub bulls and any others completely intractable were destroyed, younger bulls and cows were marked. The muster resulted in a mob of 105 big rangy steers being brought down the Arrow river to be paddocked overnight on the outskirts of Arrowtown where houses now stand.

The plan was to drive them down the Kawarau Gorge to Gibbston, up over the Coal Pit saddle and through the Nevis Valley, to rail at Garston and eventual sale at Lorneville. This alternative route to the south, though longer, avoided the tutu-lined lakeside track.

What this mob lacked in numbers they more than made up for in character. As any stockman will appreciate they were by no means docile, added to which was the fact that some genetic trait had endowed them with a fantastic spread of horn that would have delighted the old longhorn cowboys of the West, but was to make them all but impossible to get up through the loading chute and into the railway trucks.

Though these cattle had no knowledge of what fences were intended for, they'd had a couple of big days on end and remained quietly in the well-fenced paddock overnight, quickly reviving in

Calf branding. An operation that sometimes has its lively moments.

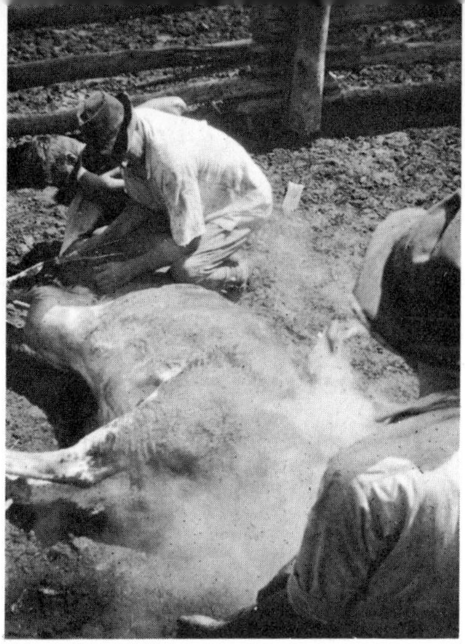

strength and spirit on the first cultivated pasture they had encountered.

Doug, who was in charge of the drive, had anticipated this revival and had issued strict instructions that all hands were to bring only two dogs and that the dogs were to be the most experienced and dependable.

Dawn was just breaking next morning as the three drovers opened the gate and rode quietly round the paddock. They had just started drifting the cattle quietly for the gate when, from a still-sleeping back street of the town, the peace of the morning was rent by the deep baying bark of a dog. Like a hound on a trail the sound rapidly drew closer and round the corner skidded a particularly boisterous young huntaway that had obviously decided he was not going to be left out of the fun.

Spotting his master behind the cattle, delighted that he had got free and had managed to catch his master up so quickly, he was announcing it to all and sundry. He went to fly the fence, misjudged his leap, hit the top barb with a bang and hung there momentarily, kicking and struggling, his barks now changed to equally loud howls of pain and annoyance. Then he dropped free and, giving vent to alternate barks and howls, rushed to join his discomfited master sitting his horse in a state of frozen apprehension. The steers took off with a rush, flattened two chains of good fence and stampeded down the road in a cloud of dust, fortunately in the right direction. They had gone two miles before Doug was able to gallop past the lead and get them steadied down.

The drive had started in a style that promised plenty of interest along the way.

At the entrance to the gorge they were spelling the mob, well back off the road, when a car pulled up and a man got out and walked across. He proved to be none other than the buyer who had failed to clinch the deal mentioned earlier. He offered to buy them on the spot, but was told to await the coming of the Skipper, who was due any minute to assist in the crossing of the first big Kawarau bridge. The Skipper duly arrived and, having been informed of the lively start, decided to accept the buyer's offer of £5 per head, delivery Cromwell — this being a much shorter drive with a good road all the way. This change of plan, whilst pleasing the drovers, was not to sway the individualists in the mob from providing continuous interest along the way.

The drovers had a sticky time getting them over the Kawarau bridge (the old one, with the sharply angled approach cut out of the bluff). Four or five meekly followed the front rider on to the bridge but, liking neither the hollow sound nor the height, came thundering back to cause a jam at the entrance. With the leaders now scratching round on the end of the bridge trying to get off it, and the mob still pressing forward from the back, a portion of the old stone safety wall bowed to the pressure and a surprised steer was left balancing on his brisket, gazing in wonder at the hungry waters of the Kawarau rushing past more than a hundred feet below. It was a tense moment: for had he gone over, the mob could well have followed him. Fortunately he regained his feet and the mob was eventually encouraged to cross.

They had just broken into a swinging trot down to the old Victoria bridge when one big steer decided that all this excitement was just too much and dropped dead, bang in the middle of the road. With the mob to be attended to there were few alternative ways of disposing of this new obstruction on the road so, with much heaving and swearing, in accord with an out-of-sight-out-of-mind policy, he was rolled over the edge to go plummeting down into the Kawarau.

They camped that night at Waitiri, without incident.

The next morning, with the front rider providing a lead and in a position to warn anyone met to take cover — for the reaction of these cattle to the sight of anyone on foot was to rush up for a closer look — they moved on down the gorge. They had the honour of being the first traffic to cross the Gentle Annie bridge, which they did in a style that confirmed the builders' confidence in their work.

When it came to going through Cromwell, the front rider went well on ahead to warn everyone to keep off the street and also to get any cars out of the way. This they did, the shopkeepers quickly closing their doors when they saw the type of cattle beginning to pour down their street. One optimistic gent had declined to move his shining new Model T and spent an anxious minute or two jumping up and down behind the protection of his shop door while a big steer, never having had the pleasure of a close inspection of

"A surprised steer was left balancing on his brisket, gazing in wonder at the hungry waters of the Kawarau . . .', over the bluff at left.

a vehicle before and not liking the look of it, decided to have it on. Hooking a three-foot horn under the radiator he flung the car up in the air then bunted it a chain down the street, walking off in disgust at this gutless thing that refused to fight back.

As the mob was making its way over the Cromwell bridge, one inquisitive steer decided to make a closer inspection of these civilised surroundings and entered the open door of a stone stable. After coursing around its interior for a while he decided it was much cooler inside and he'd just bail up for a while. How to get him out posed a real problem: there was only one door and this he jealously guarded. It soon became apparent that all was not well with this fellow, for he kept lying down.

One of the boys managed to get round behind him and get him out on the street, where he promptly collapsed in front of the pub and refused to move. By this time the rest of the mob was over the bridge and all hands were required to yard them, so a rope was brought and placed over his horns, and he was tethered to a verandah post. There he remained quietly parked while the mob was attended to.

The mob was soon yarded, with the exception of one steer, the one that had taken on the car; he refused to relinquish his freedom and independence and, jumping a fence, made his way to the top of a knob overlooking the yards and the town. There he took his stand, defying all efforts to bring him down.

With this stalemate on his hands Doug had time to remember the laggard back in the town — he hurried back to find him surrounded by a small admiring audience which included the policeman. Combined efforts failed to secure any change in the situation and, though there was no parking fee, the general feeling was that he added nothing to the appearance of the town and immediate action was required. With the assistance of a little pressure from the law the local butcher was prevailed upon to purchase him for the sum of £3, on a "where is as is" basis. The officer brought out his rifle and that problem was off Doug's hands.

But now, while the steer up the hill continued to survey all these goings-on with interest, at the same time resisting all attempts to dislodge him, the policeman led Doug down to the bridge and directed his attention to the confluence of the Kawarau and Clutha rivers. There, in the resulting whirlpool, riding high and handsome, circled a big bloated steer, the one they had hopefully disposed of back up the gorge. The officer of the law indicated in no uncertain terms that this steer was breaking several regulations, polluting the river, etc, and was to be removed forthwith if not sooner.

Now the cattle had also to be loaded or the train would go without them so, after one more abortive attempt to get the independent one down off his perch, Doug was forced to devote his attention to the quick rather than the dead, hoping at the same time this new problem would somehow sort itself out.

Eventually, after a considerable struggle, the mob was loaded and they were once more free to deal with the side issues. A check revealed the corpse still gyrating gracefully round in the river, so a quick trip was made to a dredge working further downstream, and an agreement was reached that for a fee of £5, they would bring up a boat and remove the offender.

Considerably relieved at this development, the boys were free to consider their one remaining problem, but as this one would take some working out they repaired to the pub for further discussion.

After a few rounds, who should walk in but the butcher, obviously in good humour from having a fine steer delivered on his doorstep, so to speak, for the sum of £3.

After Doug had shouted the butcher a couple of times, he tactfully broached the question of the independent one up the hill. Having had a clear view of the previous attempts being made to bring him down the butcher was inclined to drive a hard bargain, but after another round or two was convinced that all that would be required would be to take up a few quiet cattle and the steer would be bound to walk down with a minimum of trouble. In

Remarkables cattle need to be hardy. Caught in an out of season snowfall at over 4,000 feet.

addition, the butcher could not ever hope to purchase steers again as close to home as he had been able to that day, and the deal was finally concluded at £5.

It transpired afterwards that the butcher never did get that steer down alive. Despite company he refused to shift and the butcher, fearing he might see his five pounds take off for better pastures at any time, was forced to drop him on the spot, after carting up sheerlegs to hang him up on.

The only other problem exercising Doug s mind, as he turned in that night, was just how he was going to word his report of the drive to the Skipper.

The 100,000-acre Branches Station, lying thirty miles back over the 3,000-foot Skippers Saddle, takes in all the headwaters of the Shotover. It is ideal high cattle country, embracing the big valleys of the Shotover and its tributary the Polnoon.

Eighteen miles up the Shotover above the homestead, almost on the divide between the watersheds of the Shotover and Rees rivers, lie the dark waters of Lochnagar.

A most unusual body of water this, formed by a massive rock slide damming back the creek outlet from a high basin — the creek, still in its old course, now gushing forth from beneath the great rocks some distance below the level of the lake itself.

At the period of which this is written the run was in the estate of W. S. Reid and the spurs and the gully leading to the lake were clothed in a dense miniature forest of teatree and matagouri. After you have negotiated the scrub and breasted the last steep pinch, the lake is revealed — sombre waters, held fast in the tumbled rocks, scree and steep bluffs of the surrounding peaks, with the bleak saddle to the Rees at the far end.

Little sun strikes these waters — it is a wild and beautiful, if forbidding place. One's lasting impression is that here lies a corner that has changed little if at all since the Ice Age passed away.

Here lay the haunts and stronghold of the last of the wild cattle on the run — and ideally suited it was for their last stand. A labyrinth of tracks and tunnels interlaced the scrub — a rather nerve-stretching experience to climb through the silent scrub with the thought in the back of your mind that, round the next corner, you could come face to face with an over-proddy cow. If you can be "treed" up a rock, then this was just the place — as a certain then young lass, Jessie Reid (see VIII Part Two), well recalls.

These cattle which, it is thought, originated from stock grazed in the area by E. Monson of Queenstown about 1910, were a distinctive breed, predominantly white, with a massive spread of horn and some characteristics that hinted at a cross between Highland and Shorthorn cattle. Rangy in the extreme, they were as agile as deer in negotiating the rough terrain.

One little side item — in the early 1930s, Mr Monson, then well on in years and a great lover of horses, decided to visit Dublin for the Horse Show. In the process of organising his travelling arrangements, he wrote to David Reid, who had been a local boy but was at that time manager of the Bank of New Zealand in London, requesting "When the boat gets to London, meet me at the jetty." (In a way, a sound comment on the world in general from one who had spent much of his life in Queenstown Bay, and certainly a comment that could be guaranteed to deflate any officer of the great Port of London Authority had it been passed on.)

Alan Reid being ill and Jack at the war, this particular branding muster on the Branches was carried out by Nick from Mount Aurum, Sandy from Coronet Peak, and Cap from the Falls. All were experienced hill men and it proved to be a happy muster with many diverting incidents.

Sandy, however, goodnaturedly maintained it was quite a strain, what with Nick wanting to work half the night and Cap fussing about wanting to get cracking at the first light of dawn. Nick gave a short answer when asked if he had brought any sausages this time. On an earlier muster he had been buying supplies in Queenstown; spotting a great big sausage in the butcher's shop, with thoughts of many easy and tasty breakfasts he had brought it along. Big slices, nicely fried, were served up for breakfast the first morning — however, it proved to be uneatable, being full of bits

Looking across the Sixteen Mile flats to the Head of the River. J. G. Reid in foreground.

of bone and of a unique flavour. The busy butcher had omitted to inform Nick it was a dog sausage and Nick, not having caught up on such newfangled inventions, had been taken in.

From the Sixteen Mile they set out to muster the Head of the River. When passing Lochnagar they could see a small mob of the wild cattle high up on the spur to the north of the creek. Nick remarked that he had rarely seen them together in the open like that and it could be worth a go to get them out on their way back — if they were still there. Sure enough, they were still there when the men returned, and a council of war was held as to the tactics to be adopted. It was decided Sandy would hold the mob of station cattle on the flat as a decoy, while Nick did a big cast out, back and round behind them; Cap was to go to the top south edge of the scrub below the lake. The plan was to place the wild ones between them and direct the certain stampede towards the cattle on the flat.

Back up the river went Nick, and Cap worked his way quietly up through the scrub to a high point. As he reached it the wild cattle began to drop down the spur.

Travelling quietly they entered the scrub, angling across then, sighting the waiting horseman, broke abruptly downhill. Away they went full bore, slaloming down through the winding tracks in the scrub, and when the flat was reached they headed full gallop straight for the mob, where Sandy was poised to receive them. They hit the edge of the mob, bored straight through and, in total disregard of horse and dogs, shot across the river and away up the opposite face until they were stopped by an enclosing ring of bluffs.

When the mob had gone crashing down through the scrub they had put up a big red deer stag, that had assumed he was safely planted there; he was promptly dropped by Cap. Still no sign of Nick, and it was not until after the stag had been skinned that he appeared high up on the spur and began working his way down

"The steep, frowning glories of dark Loch-na-garr." *Photo: Roger Sewhoy.*

to join up. He looked a bit despondent on arrival, but brightened up when he saw the hide.

"Never saw a sign of them," said Nick, "did you?"

"Not a sign," — having him on, for obviously the cattle had scented him early on and he had not seen the performance or spotted them high up on the other side, being busy looking over the ground on his side.

When they reached the mob of station cattle and Sandy had expressed his views on wild cattle at some length, Nick got the message. It was decided that this time Nick and Sandy would hold the station mob while Cap crossed the river and climbed out to the little mob, now fidgetting about in their restraining circle of bluffs.

When Cap had made enough height he swung a huntaway up and out to one side, then set him alight. He was only a little black-and-white stumpy, but he had a heart like a lion as far as cattle were concerned, plus an excitable and annoying high-pitched cackling bark.

As he went tearing into the mob from the flank they broke round the hill — well and good, but the face they were on pitched steeper and steeper. The leader must have been hoping he was going to suddenly sprout wings, but it was not to be: instead, when the face had become almost vertical, he slipped — and away went the whole mob, rolling and bouncing down the steep face for some 400 feet in a cloud of dust.

They were tough those cattle, not one was injured in their abrupt descent. They were groggily sorting themselves out when Nick galloped over and directed them towards and into the station mob, moving the whole bunch off at a brisk pace before they had time to catch their breath.

Once these cattle had been got clear of their stamping ground they gave little trouble, apart from an occasional halfhearted break and a desire to try and hide in every patch of scrub along the way.

They kept strictly to themselves in the paddocks, eventually being sold to a buyer on the Taieri Plains. Though he hopefully kept them a long time, they never put on an ounce of fat.

Could it be that, like Byron, they fretted for the untrammelled freedom of —

— the crags that are wild and majestic,
The steep, frowning glories of dark Loch na Garr!

15

SNOW-RAKING

AMONGST the hardest and most physically demanding work undertaken by man, that of high-country snow-raking must stand as possibly one of the greatest challenges. In undertaking this a man commits himself to days, often weeks, of the most arduous physical effort under conditions of the greatest discomfort and danger.

Snow-raking, the work of locating snowbound sheep, tramping tracks through the snow to join scattered groups into larger mobs and then tramping a main track out and down on to clearer country was, in the first half of this century, carried out without the aids available today — aircraft for locating the sheep (sheep-spotting), helicopters to carry men fresh to the highest point, skis and snow-shoes, these were unknown. Strength, fitness and, above all, determination were the basic requirements, as they are still, backed by experience and stock sense.

Under the direction of the head shepherd, who must have the ability to plan the operation, assess the physical limitations of his men and carry the burden of their safety and wellbeing, these operations were carried out as successfully as conditions would permit.

If the snow has fallen heavily during a raging blizzard, the majority of the sheep may drift before it to lower and safer country; but many, seeking shelter in gullies and behind rocks into which the snow naturally drifts, are trapped and almost invariably buried. The system then is to search the places where they might have sought shelter, looking for the air-holes kept open by their breath.

If it has been a quiet snowfall they may be anywhere on the hill, in the case of a deep fall they can be buried where they stand.

As conditions deteriorate they tend to huddle into mobs and, as the snow grows deeper, crowd closer together. One flounders and goes down, then another, and the mob closes in on top. This can go on until fifty per cent or more of the mob is dead underfoot with the others roofed over by the snow, only detectable by the fouled snow round the air hole above. The men who have to liberate the living or who inadvertently fall through into the foetid air of these miniature caverns do not soon forget the experience.

The ability of hill sheep in good condition to remain alive with nothing to eat, apart from the few snow tussocks they are standing

LAMENT TO SUMMER

Gone the whisper of wind in the tussock,
The cheerful music of stream and rill.
Gone the rasping song of the cricket —
The call of duck and pippet's trill.

Gone the standards of speargrass blooming,
Rank on rank on hillside bright.
Gone the elusive scent of annise —
Shy edelweiss, gentian and lily white.

Only the silence of snowbound reaches,
Fast in the grip of winter's thrall.
Only snow banners, from peaks a'flying —
Silence, cold — and a kea's call.

Snow-raking — the high-country equivalent of search and rescue.

on, for periods of up to a month, is amazing. Each mob snowed in presents a different set of problems, according to the location and snow conditions. With some it is sufficient to tramp the track up to them and they will take it with little urging. Others, seemingly befuddled by their long captivity, will flounder off in all directions and will have to be carried bodily back to the track, usually covered in great balls of snow which have to be cut off before they can walk. Trapped mobs become easy prey to the keas, too. Being unable to escape, they become living larders, and those that have not already died from blood poisoning have to be despatched by the knife.

They were snow-raking and had climbed that day for the Two Mile hut. Six of them, each carried in his pack an allotted portion of the total rations for four days, as well as a couple of pairs of dry socks and a spare shirt.

There was no clear ground. It had been snow-straps from the word go and as they climbed higher the snow became deeper and the drifts worse. As they turned the spur at 5000 feet into the Two Mile basin it was up to their hips, two feet six inches to three feet on the open ground. A deep loose powder on the south faces, and the north faces covered by a crust that was just not quite strong enough to carry a man's weight, jagging and bruising at their hips as they took turns breaking the track. Walking through deep powder snow is like walking in the dark, all one sees is the smooth white surface, underneath there can be anything.

The higher gullies were completely drifted in, level across, fifty to eighty feet of soft powder in some, into which a man could disappear as if he had stepped into a huge tub of suds. They crossed the gullies at the narrowest and shallowest parts, testing the depth with their hillsticks. Even so, the man in the lead was caught out several times. Testing the depth in front he would push out across. Not too bad, then — *whoosh* — straight down. He had walked off the end of a big rock or small bluff — all rather amusing for those still on top, but a heart-stopping experience for the one concerned, as the icy blue-white powder poured over his head, down his neck, into his ears and nose, not knowing how deep it actually was. Floundering and flailing he would hit bottom, look up at grinning faces and thankfully grab the extended hillstick and be hauled out to brush off and shake out the fine powder as best he could. The one who laughed most or made some witty remark found himself next on the list to lead. As they came into the steeper country they chose their track more carefully. Here, in the chutes, underneath the powder lay an earlier fall, rock-hard, from which the last fall would avalanche at the drop of a hat.

Up till now they had been climbing in the brilliant sunshine of a clear calm day, which had been accentuated by the sun's reflection from the still waters of the lake far below. The warmth had been pleasant, but the glare from the snow was, after a time, an almost living bludgeoning force. Despite the use of burnt cork to nose and eye-sockets, and eye-masks made from a bit of sweaty old black hatband with little slits to see through, the fear of snowblindness made them keep their gaze, when possible, on a dog or the back of the man in front. Now, as they turned into the shadow of the valley they had relief from the glare, but a temperature drop soon caused coats to be put on and old socks to be drawn over the hands. A stiff breeze was blowing down the valley, whirling the fine powder into their faces, coating their clothes and the dogs' coats. The surface of the snow was obscured by the drifting veil, making it difficult to see ahead.

They made their way down to the creek to cross to the easier sunny side, sliding and floundering down the short spur, the snow tussock underneath as slippery as ice. The creek was completely buried, only traceable by the gurgling and splashing from beneath. After much prospecting with hillsticks they found a bridge of older snow and stepped gingerly across, back into the sunshine of the opposite side.

Not far now to the hut. From the top of the next rise the Two Mile basin was in full view and the head shepherd said "There's the hut." The ones who knew the country chuckled as the new hands searched the snow-covered basin in vain, and laughed at their expressions as the head shepherd pointed under his feet — they were standing on the roof. It was almost completely drifted

The Two Mile hut. In the episode here described it was almost completely buried in the snow.

over, only a bit of the bottom side roof and wall showing. They scooped the snow away from the door and the window and removed the pile from inside the chimney. It had taken them six hours to make what was normally a two-hour climb.

A couple of hours of daylight remained and as the head shepherd was anxious to locate some of the mobs of sheep that day, in this one of their favourite haunts, he detailed the youngest member to clear up the hut and get some food ready while the others pressed on. On the way up they had seen and passed close to several small mobs, standing marooned and helpless. Up here they could expect to find a greater concentration of sheep.

In the earliest days of the grazing of the high country it had been the practice to follow, in some degree, a herding policy somewhat similar to that of the older lands. The present system of free-range grazing was yet to evolve.

This hut had been built by John Cockburn, one of the early herders, who had used it as his base during the summer months.

It was eight feet by twelve, a standard size for many of the old mountain huts, which started often as merely a skeleton of walls and ends, to be roofed by a tent fly as required. Eventually these were permanently roofed with corrugated iron and attained the status of a hut. Built of stone and mud, the side walls are about four feet high and eighteen inches thick with, on one side, a small window eighteen inches square, the other side wall being set into the slope of the hill almost to the roof line.

One end is solid, the other being broken by a narrow door and a fireplace and chimney. The floor is of paving stones, through which, in wet weather, miniature creeks trickle, seepage from the hill behind. One half of the hut, opposite the fireplace, is taken up with a field bunk, merely a raised platform of planks covered with cut snowgrass. The rafters had first been covered with sacking, then snowgrass placed between that and the iron in an attempt at insulation. On a cross strut were hung the eight big rough blankets covered by a rag rug, or wagga. Below the window a large box, tin-lined, with a door, served as cupboard and table. The tin lining was protection against the fieldmice who, for some unknown reason, possibly the snowgrass seed, are so ill advised as to live at this altitude. The opposite side was taken up by a half bag of coal, a pile of peat blocks, some wood and a box for the camp oven and pots. The fireplace, with its blackened hooks for pots, must be one of the worst ever built — cook on it, yes; throw out any heat, no; but it smoked inside, superbly.

Where the roof met the walls were many small gaps which, despite plugging with a variety of materials, in winter let in a fine mist of drifting snow. It was always reckoned that at the Two Mile hut one did not need to blow out the candle, but simply snapped off the flame.

In all, a building which, today, would be described as having "a lot of character".

The boy watched the others tramp on into the basin — he would rather have been with them than acting cook, but knew how important a hot meal would be. He turned back into the hut.

First the fire, a good big one to dry out the chimney and leave a good steady heat for the cooking. Water from the creek took some getting as it was a case of "first find your creek". Then peel the potatoes and onions to go with the "bully" and get them on, along with some rice with plenty of raisins. Once these were on he turned his attention to the bunk, which was liberally coated with fine powder snow; he would have to get that out before it got rubbed in, so he carefully removed the first layer of snow tussock and the snow.

The blankets, covered by the wagga, were as dry as most mountain hut blankets can hope to be. These he spread out on the bunk, and now everything was ready for the others' return.

By this time the sun had gone and immediately the temperature dropped — from way up in the basin he heard the sound of dogs and wondered what the others were doing.

It was just getting dark when the others came back down their track, their footsteps making that peculiar squeaking crunch of dry snow at very low temperatures. In the small hard-packed area at the door they dusted off the snow and removed their snow-straps before crowding inside. The dogs were not tied up and they soon set to work burrowing into the snow for the night.

After tea and with the billy filled up ready for the morning, they soon turned in for, although the fire had been well built up with dried peat blocks and the light of two candles gave an illusion of cosiness, the actual temperature was quite arctic.

Boots only were removed, the old hands placing these under their coats, which served as pillows, for experience with this hut before had taught them they would be frozen solid in the morning, unless.

They settled in for the night on the field bunk, like a row of sardines, the cold creeping up from below the damp clothes, twisting and turning for hours till in desperation they shook up the fire and made a brew of tea. Then back to the bunk — the dogs, startled by an avalanche in the bluffs across the creek, set up a racket.

The stars glittered above the white peaks and the silence fairly creaked with the cold. So in fitful dozing the night passed.

With the faint increase in light from the tiny window (the stars had barely begun to fade) the head shepherd squirmed out from beneath the blankets and lit a candle. Pulling on his boots he turned to the fire and, sweeping aside the grey peat ash, uncovered the glowing coals beneath. Some sticks of wood, a handful or two of coal, more wood, and soon the hut was filled with the glow and crackle of a good fire and the promising smell of chops sizzling and spitting in the camp oven. By the time the others had struggled out, rubbing circulation back into numbed hips (for snow tussock cannot be described as resilient bedding) and washed in a handful or so of powder snow outside, the toast was done and breakfast soon disposed of in the usual restrained silence.

With the cutting of the lunches came the head shepherd's instructions for the day: three to go back up into the basin and bring out the sheep already mobbed up, then tramp them out and down the creek to the ridges swept clear by the wind, at the same time joining up any mobs from the faces above, as far round as the Big Spur saddle.

Himself, another and the boy, to climb up to and through the Two Mile saddle and make a start on clearing the headwater basins of the Big Spur. All hands to meet and camp the night at the Upper Staircase hut.

The day had dawned bright and clear — the snow-covered spur rose above them to disappear into the saddle, no sun had reached

"The system is to search the places where the sheep may have found shelter, looking for the air-holes kept open by their breath." Nowadays aircraft are of immense help in sheep-spotting.

the basin as yet, only the main peaks catching the first bright rays. The temperature was so low, breathing was painful and expelled breath froze on the face of man and dog alike as they began the climb up the ridge partially swept by wind. They came then to the heavy snow below the saddle. Waist-deep, soft with a stiff crust on top, it took all the strength of the one in the lead to force a way up the steep face. By now the sun was fully up and the blinding reflection from the snow forced them to put on the makeshift masks. Eventually they broke through to the windswept saddle at just under the 6000-foot level.

Before them lay a silent white world, apparently devoid of life.

Below, the valley wound away from the top basins, to pass from sight several miles away. On the left lay the very steep face of the Neutrals Spur, impassably dangerous under these conditions. On their right, the back of the Big Spur, with its row of high bluffs through the middle that had claimed the life of a shepherd in the not too distant past.

They sat in the snow and had a bite while the head shepherd scanned the valley with fieldglasses, endeavouring to locate any sheep and plan the attempt to get them out — he spotted several mobs, mostly twenties or thirties. They would only be able to take those below the bluffs today — for those above the bluffs another day would be required. It was a daunting thought to contemplate the long tramp down the valley, let alone the job of collecting the mobs.

A bitter wind began lifting the snow in the saddle to send it whispering down into the basins, and they rose to go.

They had not gone far, sliding swiftly down into the first small basin, when four keas rose screeching from a jumble of rocks ahead, circled close over them and flew off to the far bluffs. Floundering in and out through the boulders they found the attraction — about fifteen sheep had been snowed-in there. Four were dead, three were lying resigned and as good as dead, with great holes torn through their backs, two more were standing with heads hanging, their eyes dull and glazed in the final stages of death from blood poisoning. The remainder all carried kea flags — this then was the first mob for the day.

With a burning frustrated hatred of the killers, they set to and butchered those beyond saving and began tramping a track clear of the rocks and on down the valley to a spot below another mob. From here the head shepherd sent the others back to get the sheep while he began breaking another track up to the mob above.

They broke the last few yards through into the trampled stained area where the survivors stood, imprisoned in their walls of snow, then climbed round behind them. They needed no urging to take the track. With a bound a big wether took the lead and they went floundering off down, to bring up bogged where the track ended.

The man and the boy went on down and began improving the head shepherd's initial track and that mob was soon floundering down. They created miniature avalanches in the steep places, ending up half buried. It was necessary to get to them quickly when this happened and start unravelling the kicking, snow-covered mass — no time to be worrying about freezing fingers, as sheep covered in balls of snow were hauled free and slid into the track ahead. Eventually the mob reached those below and a track was tramped through to below the next to be rescued.

So the day wore on, and the main mob gradually grew in size.

But now conditions were deteriorating. The wind had grown much stronger and they could only tramp a track for a short distance before it was filled again with drifting snow. To make matters worse it had clouded over and snow began to fall — this created the dreaded white-out, a condition in which it is still possible to see, but not to be able to tell where the surface of the snow is or any details — it is only too easy to step off the end of a bluff or into a sharp hollow.

As he must allow time for them to reach the safety of the hut before it became completely dark, the head shepherd decided to leave the mob till next day and to press on, with the faint hope that the mob would follow their track at least for some of the way.

They made better time on the flats, floundering along in the growing dusk with the snow sifting down.

They rounded the last spur and saw the hut — a drift of smoke coming from the chimney and a glow of light from the window.

The others had been able to make good time along windswept ground on the Front Face and had got in before them.

This hut was just unlined corrugated iron, but sported a wood floor and a good fireplace, which was just as well, considering the insulating properties of the rest.

They were grouped round the fire after their meal, some drying socks, others rubbing old fat into their boots, when the dogs began to bark and they heard someone step up to the door. They looked at each other in amazement, one of them threw the door open, to turn and say "Good God, it's the Skipper!"

As he stood outside in the darkness and swirling snow, kicking the snow from his boots against a rock, he called the head shepherd out.

He had ridden sixteen miles, climbed five miles and 5000 feet, breaking his own track through deep drifted snow, to bring the head shepherd word of his father's death.

That early snow, so dreaded in the high country in the fall, had caught the sheep still out on their summer blocks.

It took six weeks of snow-raking to save the major part of the flock — six weeks of grinding toil. Men became so tired, day merged into day, but they would think of the climb the Skipper had made, and kept going.

Spring at the Upper Staircase hut. Not a sophisticated residential unit, but a blessedly welcome sight to exhausted men.

16

LAMB-MARKING

For lamb, as Man,
The seasons swing,
Yet summer, winter,
Needs follow spring.
The idyllic life of carefree days,
Is bound by Fate to other ways.

OF all the jobs in the annual round of work on a high-country property that of lamb-marking provides the greatest scope for the occurrence of unforeseen events — and that would be putting it mildly.

All high-country men must carry with them rueful memories of those days when the best-laid plans have gone agley. For, of all breeds of sheep, the Merino commonly bred in the hills must be among the wildest.

If the runholder is to gain the maximum benefit from the great areas of sparse mountain grasslands under his control, a breed of sheep must be selected which has the physical qualities to enable it to cover great distances; it must have no inclination to mob up and overgraze any particular area; it must be able to provide an economic return and at the same time be capable of withstanding the severe climatic conditions. These qualities the Merino has in full measure.

In physical structure it is neat-footed, leggy, and comparatively light in bone, as is another of the great free-ranging animals, the red deer. It tends to carry its reserves of fat in the most protected and readily accessible position, inside. It is capable of producing a heavy fleece of high quality fine wool under conditions in which other breeds of sheep would be in danger of starving.

It is independent by nature, preferring to graze alone or in the company of a few others rather than with a large flock. It is neither timid nor shy, yet resents the intrusion of human beings or dogs into its area and will seek to escape from their presence.

All these qualities make it an ideal breed for use in most of New Zealand's high country. It is free-mustering and at home in the wild environment. The fact that its special qualities are genetically strong makes it an ideal sheep for cross-breeding where, the particular block of country being suitable, it is desired to introduce other qualities without diffusing the Merino's desirable characteristics.

During lambing the Merino ewe does best under conditions approximating to her preferred environment — a free range and as little disturbance as possible. Where, with other breeds in different environments, the daily attention of a lambing shepherd can have beneficial effects on the success of the lambing, in the case of the Merino, even with the best of intentions, it may well have the opposite effect.

For the benefit of the uninitiated readers it will be necessary to go into some detail, to enable them to appreciate the problems inseparable from the handling of a large mob of Merino ewes and lambs.

Though the yarding of any mob of ewes and lambs presents more or less similar problems, a mob of Merinos off a high-country tussock block can often bring these same problems to such a pitch of intensity that strong men turn pale while others will jump on their own hats. Success will depend on meticulous planning, a knowledge of the normal pattern of movement of the sheep on the particular block, split-second timing, sheep sense and — luck, buckets of it.

The major considerations involved are these: the site for the yards should, if possible, have the following attributes. It should be located in an area which the ewes are used to traversing, against a long straight fenceline, near the top of a gently sloping piece of unbroken clear ground, with the slope rising up to the final pens; there should be no major obstructions within a mile of the yards (such as steep gullies or big patches of scrub through which the forming mob has to pass); the yards should head into the wind and be designed to easily accept and contain the mob; there should be as big a wing to the yards as material and time will allow; there should be no movement round the yards as the mob approaches, nor should horses, dogs or vehicles be left anywhere near; the gap through which the mob enters the main yard should be as wide as possible and capable of instant closure; the musterers should be experienced and should have good dogs, and all pups or any white dog should be left at home that day.

As the first active leaders of the mob arrive they should be very gently checked from either entering or bypassing the yards. They should on no account be turned back, but encouraged to settle quietly in the ample space below the yards where the mob will eventually form. This undisturbed formation of the mob is critical.

The strings of ewes and lambs should drift in quietly, with no mismothering, to join the mob forming below the yards. As the tail of the mob closes up the sheep must be able to see the withdrawal, well out to the wing side, of the person who has checked

Yarded wethers. "Of all breeds of sheep, the Merino bred in the hills must be the wildest."

them. As he withdraws the leaders will sense the opening and move uphill, into wind and deep into the yards, drawing the main mob smoothly after them.

This is the critical point in the whole day's operation, for now mismothering cannot be avoided. Up till now the mob has been handled as smoothly and quietly as possible — but now the aim is to stampede them into the yards before they get a chance to realise they are trapped and try to get back out. The timing of this sudden burst of force is absolutely critical, a few seconds either way can spell the difference between success or ignominious failure.

At a signal from the head shepherd all hell breaks loose, wildly barking dogs, shouting and screaming men, flogging coats — anything and everything that will accomplish this final rapid stampede deep into the welcoming arms of the wings and yards. A portion of the wing is detached and swung, just as quickly as it can be done, round behind the shepherds doing the forcing at the back of the mob. This effectively contains any sheep or lambs that may break back past the first line of attackers. As the tail of the mob enters the yard proper, the closing portion of the wing is brought right up, the stakes or waratahs are driven — the mob is yarded.

Now, and only now, attention is given to the few lambs that may have got away — a suitable number of ewes are drafted off and allowed to mix with the escaped lambs. This little mob is then directed back into the main mob through a closely guarded gap, which is immediately closed.

Dogs are tied to stakes inside and along the back of the yard and at any place where the mob may be likely to create pressure on the light temporary yards — the major and most nerve-racking part of the day has been successfully accomplished and the work of marking the thousand or more lambs can be commenced.

"What a pipe-dream!" any old hand who may read this will scoff — and he will be entirely correct. For seldom if ever is it that such an ideal site is available, or that the sequence of events will proceed as outlined.

The mob may arrive in a mismothered state, from a hangup at a gully well back in the muster. The leaders may be just plain difficult and go bolting back on the approaching strings, mismothering and upsetting everything as they go. It may come up a wild blustery day with the wind from the wrong direction. The mob may have to traverse a steep slope, which will siphon off, as through a sieve, the majority of the lambs to the top side; or the dogs, who are quick to sense any tension or worry on the part of their masters, may become excited and over-run and check the lead or, more fatally, start the mob ringing. Equally disastrous, a moment's carelessness on the part of a dog may serve to detach a wing of the main mob that is rich in lambs. For any of a dozen reasons the mob may have to be forced too soon.

Any of these can lead to the dreaded lamb-break, hundreds of lambs go slicing off across hill and dale, going like the wind, their long tails flogging them on. The frenzied mothers try to break away after them and general pandemonium breaks out — frantic shepherds, trying to contain the mob and make their dogs hear them above the din of hundreds of baa-ing sheep, have their hands more than full and struggle to get what they can of the main mob into the yards. It may take hours of careful and frustrating work to get the delinquents back with the main mob.

It can be taken as read that at least some of these occurrences will take place during a marking muster. Once the mob begins to form, success or failure are balanced on a knife edge.

It had been a good spring and the Merino ewes grazing untrammelled over the abundant expanse of their lambing block were content.

Not for them, this spring, the ceaseless search for food to meet the increasing demands of the growing lambs they carried, with successions of late blizzards interspersed with frosty nights to hold back the pasture growth; this spring they ranged far and wide only to satisfy their inborn instinct to travel. Warmed by the sun of the

Keystones of a nation's prosperity.

lengthening days, the north-facing slopes of their favourite night camps remained warm for much of the night. The ample feed enabled them to remain on their camps with no need to go foraging in the dark. Undisturbed as they were, time too could be spent chewing the cud and resting on the high points of the spurs during the day. It was ideal lambing weather and the number of lambs throughout the flock gradually increased.

The little ewe had become more conscious of the weight of her unborn lamb these last few days, it had become increasingly difficult to climb out over the spurs to the distant plateaus where the feed was sweet. As the sun rose on the camp that morning she had felt restless and ill at ease with her companions. Instead of drifting off with the others on their leisurely trek to the distant grazing, she turned to seek the solitude of the higher slopes. Sorely troubled, she wound her way upward with many a pause for breath, the sweet tufts of grass on the way of no interest to her. Finally she reached a series of small basins in which there were widely scattered pairs of mothers and recently born lambs.

As she passed some of these she was impelled to inspect the lambs closely. Sniffing at them she sought to find a scent that appealed and attracted her; none had it, and their mothers resented her intrusion, bunting her sharply away.

The hours passed as she drifted restlessly on, now completely dominated by the instincts wise old Nature had decreed — she came on a tiny basin, just a fold in the hill, and was pushing through the soft, gently waving silver tussock, when suddenly she felt more at ease, she grazed for a short time then rested.

The shadows were lengthening when the activity of her lamb within and the instincts that now controlled her, drove her once more to her feet. She shifted to a place where there were a few feet of clear ground and attempted to lie down again, but it did not seem quite right. She scraped at the ground with her front feet, working round and round until she had formed a comfortable little hollow. In this she lay down, with no thought or inclination to return to the usual night camp far below — as dusk was settling she became racked by the first spasms of birth.

The first stars were bright in the sky when her lamb was at last born and she lay, enveloped in vast relief from the uncontrollable spasms — but only for a few seconds. She was startled by the lamb sharply shaking its head, she leapt to her feet and cautiously inspected the new arrival; as it struggled to get to its feet she began to lick it dry. It was soon standing, propped on shaky legs, staggering under her attentions. Within a minute it began staggering along her side, instinct driving it to seek the promise of life from its mother's warm nipples — it suckled and presently lay down, the ewe settling down beside it. Many times that night she rose to suckle the lamb, for it did not rest for long periods in the chill night air.

In the first welcome rays of the rising sun the ewe rested, the lamb close at her side — what a beautiful lamb it was, a sturdy ram lamb, white as the driven snow. Almost unable to take her eyes off it, she periodically nuzzled and sniffed at it, satisfying herself that all was well.

Below her lay the great sweep of face and spur, the tussocks gold and silver, the bluffs and rock outcrops a warming grey in the rising sun. Here and there the distant forms of ewes were moving slowly, just commencing their grazing for the day; following them, in bursts of activity, were the white specks of their lambs. Occasionally the distant burbling baa of a ewe, feeding and calling her lamb at the same time, could be faintly heard, to be answered by the fainter bleat of the erring lamb — a deep, living peace lay over the hill.

The pangs of hunger now drove her to graze, but no more than a few feet from where the lamb slept. At its slightest movement she was back to nuzzle it to its feet, to kneel once again to suckle, as only the lamb has the grace to do. By afternoon it was following her as she fed, occasionally falling asleep in the warmth of a sheltering tussock.

As the ewe grazed on she kept a close watch on the spot where the lamb was sleeping, and it was well that she did so — for, with a swish of wings, a black-backed gull swept down to settle on the tussock where the lamb slept. With a whistle of alarm the ewe raced back through the tussock, charging the gull. It rose with powerful wingbeats to glide off round the face in search of easier prey — a few seconds longer and the lamb, unable to move quickly at this early age, would have been minus eye or tongue.

Shortly afterwards she received another scare — a shepherd on horseback appeared abruptly over the ridge and rode right into her hollow, two dreaded dogs close behind. With a snort she started to race away, suddenly remembered the lamb and raced back. One of the dogs had stopped to sniff at the sleeping lamb when she arrived like a bolt from the blue; head-down she charged, unmindful of her fear, her only concern her lamb.

The dog, taken unawares, was bowled over and would have taken her by the nose had not the laughing shepherd called him away. He trotted after his master with the sheepish look of one who has been caught with his pants down, while the little ewe stood beside her lamb stamping her front feet and whistling defiance.

Badly shaken by the encounter she realised she could no longer continue the idyll of their solitary state and must seek the greater safety of numbers; she moved steadily off to join a small group of ewes lower down, pausing occasionally to wait for the lamb, gamely floundering through the tussocks in her wake.

The days passed, with the lambs rapidly gaining strength and mobility. In the evenings they would form into a group, chasing each other in follow-the-leader games, up and down through the big rocks. The ewes, tapering off their evening grazing close at hand, watched these capers with indulgent eyes, the younger ewes occasionally voicing their concern if the dancing band of lambs got too far away.

As dusk settled, they would break off in ones and twos to dance back to their mothers — the ram lamb was always amongst the last to return. Now well grown and strong he would come bouncing back to suckle, boisterously bunting her udder and nearly lifting her off the ground, an indignity she accepted with proud indulgence.

Early one morning there came the sound of barking dogs from way round the face. Made anxious by the repeated sound, which drew steadily closer, the little band of ewes called their now strong and active lambs and, forming a string, with more or less obedient lambs at heel, sought to move away from this disturbance.

Above and below them they could see other bands stringing away. The marking muster was upon them.

17

SHEARING MUSTER

In rock-hoppers' slippers
By Pannells made
We'll tramp the peak
Bluff, chutes and glade.
With horseshoe nails at heel and toe:
They'll hold on icy-rock and snow.

THE shearing muster, from the point of view of the young shepherd, was the most satisfying — untroubled as he was, as yet, by the responsibilities and implications of the past year's work as dictated and revealed by the shearing tallies — that moment of truth for all high-country stations.

To him it signified the start of a new season's mustering, with release once more from the constriction of work on the flats. The relative tedium of the winter's feeding-out and yard work, the weary and often discouraging lambing beats, the lamb-marking and unending succession of tedious, but nonetheless important, odd jobs — all were now behind and the way back to the hills lay open.

The opportunity to test the worth of the many hours of training put into the young heading and huntaway pups; new country to be seen on the other stations in the run; new faces in the mustering gang; and the chance to see the work of the legendary dogs of a new musterer famed in the dog-trial world; all these now combined with the increasing feeling of urgency as all work on the station centred on preparations for the muster and shearing. He had, with the knowledge of the months of mustering ahead, difficulty in hiding his excitement behind the guise of the blasé acceptance of an experienced musterer.

The shearing muster was the easiest muster — some of the true summer blocks were not yet stocked, yet had to be gone over to pick up any small mobs which had drifted back up, compelled even at this early time of the season to seek out their favourite haunts. On the other blocks the edge of the spring growth had not yet crept back up to the higher levels where, although largely freed from their winter burden of snow, the grass and tussock still lay flattened and browned. The sheep were more concentrated, content to graze the fresh growth of the warm lower slopes, excepting those who, over-endowed with the Merino's desire for high places,

Shearing muster. The tail-end of a mob of 4,000 coming off the Staircase.

grazed even to the edge of the receding snowfields, content to accept the freedom of the high places as compensation for the harsh grazing.

With the edge of summer making itself felt, the weather was usually good for mustering at this time of year; crisp mornings with plenty of heat in the sun later in the day. The day's musters were generally shorter, too, many of the high back basins being declared clear by the head shepherd.

The top and outside men did not have to travel as high or as wide as in other musters, often being able to use the hard-packed but rapidly receding old snowfields which, smooth as a surfaced road, covered the jumbled rock and scree of the high slopes.

Top beat was still no sinecure, however. The firm and easily travelled snowfields usually meant danger in the chutes ahead. If the day should chance to be over-mild, almost any exertion in the way of skirting the snowfields, either above or below, is preferable to facing up to the nightmarish struggle to cross the porridgy mass; snow-straps are essential and, with all surface indications long since erased, anything may lie beneath the smoothly deceptive surface. Anyone who has floundered for any length of time across a steep field of shattered rock, covered with two to three feet of almost liquid snow, is in no hurry to repeat the experience.

Good as the weather usually is at low altitudes at this time of year, winter has not moved so far on its journey to the north that it does not often reach back and scuff a heel across the last of the land within reach. That the high mountains offer no guarantee of stable weather, often changing with almost unbelievable rapidity and extent, was the lesson the boy was to learn, in no uncertain manner, on this particular day.

The Forks, that massive spur of land lying within the two arms of Wye Creek, stretches back seven miles to the main divide. For almost four-fifths of its length the top varies little from the 7000-foot level then plunges steeply to the fork of the Wye at 1500 feet — this was the day's muster.

Three men for the Back Face — up through Hawke's Pass, that narrow ledge cleaving a giant bluff. Four hundred feet above the roaring waters of the Right Hand Branch and only three feet wide in places, it was no place for a careless stumble in darkness or a high wind — then on between the towering bluffs to the level valley floor and the last steep climb to Lake Hope.

Three, including the boy and led by the head shepherd for the Front Face, climbed up the Steep Pinch, as good a pipe-opener as was ever found for a long day, on through the bush to the long walk up the valley to the Bridal Veil Falls streaming their tracery of spray below the grim face of the Double Cone.

It was not a pleasant climb through the bush that morning. Outside the bush the hills rose steeply, their peaks dark and withdrawn against the first reluctant retreat of night before the tide of day; but here in the bush, darkness almost, but not quite, still reigned — a darkness filled with spectral black twisted trunks, straining to hold aloft their protective canopy, soughing and rattling beneath the onslaughts of a cold boisterous wind.

Underfoot, the deep leafmould, still damp from winter snows, provided treacherous footing on the steep slopes, the track in many places lost beneath a maze of shattered trunks and branches where whole battalions of trees had been vanquished and crushed by winter storms.

The sound of wind, the protesting screech of boughs grown too close together in their struggle up to light, and the clatter of the occasional falling branch were but light accompaniment to the threatening rumble and roar of the creek as, rising and falling in waves of sound on the eddies of wind, it filled and dominated the darkness.

A dark, repelling place, seemingly abandoned to the elemental powers of Nature — yet here dwells one of Nature's miracles: occasionally, and amazingly, within the lull of creek and wind, there was heard, like sparklets of sunshine from the darkness, the delicate, barely audible, yet cheerful twitter of the tiny rifleman — his day begun, even at this early hour and in those conditions.

That such delicate wisps of cheerful life should choose a habitat subject to the unrestrained powers of Nature and go their way in bright unconcern of conditions that expel all other, so-called higher life forms, is surely one of Nature's wonders.

The tiny rifleman — "one of Nature's miracles;"

Dawn was pouring in over the eastern peaks as they reached the thinning outer bush. Here they paused for a breather beside the ruins of an old hut.

A touch of mystery surrounds this reminder of the past. It stands on the flat shoulder of a small spur close above the creek. When built it would have stood at the outer upper edge of the bush with a spectacular view of the valley sweeping down to the lake, the serried peaks behind stretching into the distance. Now it is surrounded by bush, a shadow of its former self and, in all this great tract of mountain lands, is practically the only visible record of human contact and association down the years.

Who built it and for what purpose are not known. It provides a regular topic of speculation at this usual stopping place.

It was probably built and used as a summer camp when this country was first stocked and the "old-country" system of herding was still practised. Whatever the reason, it had been well built, with low walls, built log-cabin style from the beech at hand and fastened with carefully shaped wooden pins. The chimney was also of logs, plastered with mud. The roof had long since rotted away and had probably been thatched with snowgrass or simply covered by a tent fly.

No great imagination was needed to visualise this shepherd of days gone by — after the long calm day on the hill and with the flock now settled content in the basin above; sitting smoking outside he would be, in the warm summer's gloaming — a pot of stew bubbling comfortably in the hut and the sweet-smelling smoke of the beech fire drifting lazily away.

Beside him, his two dogs would be stretched out to well-earned rest in the tussock; and below, the great sweep of valley in shadow and the distant silver sheen of the lake. Above and far to the west only the highest peaks would hold the last rosy glow of the setting sun.

Who he was and from what distant land he came, no one knows. But of one thing we can be certain, here he would have known moments of complete peace.

Above the bush now and sidling round through the tongue of an old rockfall, the head shepherd followed faint marks from rock to rock — worn by hoof and hobnail down the years, they indicated the only easy track through what was a natural maze.

Jumbled rock piled on jumbled rock, with an over-liberal dressing of old-man speargrass and the final edge of the tea-tree scrub. Almost continuous decisions of direction were required to stick to the easy way — many's the time the leader here has strayed from the track to end up clambering over rocks and cursing the speargrass, having to swallow his pride by seeing — either above or below — one of the old dogs, tail waving gaily and followed by a long string of other dogs, trotting smoothly along, with the occasional curious glance at those humans who seemed to prefer floundering about in the rocks to a good trail. He was a wise man who, especially in the dark, casually paused to let one of his old dogs ahead in this place. With the white tip of the old tail waving like a beacon in the dark and an obvious appreciation of what was required, he will assume responsibility of leadership and trot steadily along on faint trails long since lost to human eyesight.

Across the creek, a short steep climb and the familiar valley lay before them — the towering bluffs of the back of the Lumber Box on their left, the slightly easier slopes of the Forks country on their right and, miles away, at the head of the gently rising valley floor, rose the sharply outlined mass of the Double Cone. Edged now by the rays of the rising sun, it thrust from its base of bluff and scree the protruding tip of some gigantic obsidian blade. The dark faces slashed with ice and the rock fields at its base buried deep in snow — it stood aloof, in unquestioned supremacy of power, as if receiving due homage from the lesser peaks.

How different lay the valley now, from what it had been in the kindly days of summer.

At this altitude, spring had still much to do to wrest the country clear from winter's grip — the dark faces of the Lumber Box carried much snow, with the pale gleam of ice on the bluffs. Opposite, the tops and high boulder fields of the Forks also remained under a heavy mantle.

Below the snowline and right up the valley floor, a broad expanse of snow tussock — not yet stirred to life, it lay bleached and brown and flattened, only recently released from its burden of snow.

Gullies, still filled flat with snow, ran down to the creek and, at the foot of every chute off the dark face, right up the valley, lay huge fans of snow — mute evidence of winter's dreaded powder-snow avalanches. Oh, so quietly do they come, this type, not with the crash and rumble of sheet or soft-snow avalanche, but every bit as devastating and as deadly in their power.

The conditions suitable to the occurrence of powder-snow avalanches are normally found in midwinter. The succession of cold fronts, sweeping up from their breeding ground in the deep south, have already deposited a heavy coating of snow over the mountain lands. Then comes a slow-moving front — formed by some quirk of Nature and spawned of the very heart of winter — the temperature within this front is very much lower than that of those preceding it.

The minute particles of moisture, in the warmer air absorbed by the front, are instantly frozen. They do not have time to coagulate,

The familiar Wye valley on — "a kindly day of summer".

to freeze together into the relatively robust and intricate patterns of the ordinary snowflake that falls at lower levels. The front becomes a suspended mass of tiny ice crystals, similar to those forming the "filly-scarts and mares' tails" of the high cirrus — with time, in response to gravitational laws, they form groups and settle in a soft mantle on the mountain lands within their tolerant temperature range.

So infinitely light and delicate are these crystals that the ensuing mantle is almost 75 per cent air — it will lie in this state until the temperature rises slightly, then a several-foot depth will settle into a few inches of firm snow.

So light and airy is it, one can walk through a depth of several feet almost unaware it is there — the merest breath of air is sufficient to set it moving. A wind will send it flying in great clouds, up from one face of a mountain, over the top, to settle gently on the lee slopes. Here it can build up to great depths, in a delicate balance between friction and gravity — there would be less friction between these snow crystals than between particles of graphite.

Should the balance move more to the side of gravity, or be upset by the passage of a man, or a hare, or even by the vibrations of sound — then comes the avalanche.

It starts with a sibilant hiss, and movement is seen over the whole snow-covered basin. Hundreds, even thousands of tons of snow are on the move.

Almost instantly it reaches a terminal velocity of well over a hundred miles an hour — any obstruction, such as a small bluff or the constriction of the chutes below, seems to trigger off shockwaves, almost explosions, within the mass — great billowing clouds of snow are thrown out and ahead — each shockwave seems to accelerate the speed of the main mass and, in seconds, it has hit the valley floor.

Here the final shockwave is thrown out — a blast of air and snow, travelling at fantastic speed, fans out across the valley and up the opposite face. Then comes the noise — not the grinding roar of other avalanches but a great rumbling boom of displaced air.

Snow continues to slide quietly down the chute to join the heap in the valley — in minutes it has set into a firm mass and eventually will become almost ice hard.

Here where they stood, there had been a really big avalanche. Plunging down from the long face of the Forks, it had swept right over the creek to the other side of the valley — the snow had melted but had left behind thousands of tons of rock and mud. Beneath this for a hundred yards ran the creek — sometime in the future it could block here, dam back, then break out and devastate the lower valley.

It was a land still half paralysed by winter's cold, of drab and muted colours, with only a tentative murmur from the smaller creeks — the sharp sterile smell of snow hung over all.

They pushed on up the valley, the keen wind at their backs, with no inclination to linger in such surroundings. However, after half an hour, the sun, with what seemed like a sudden leap, surmounted the tops, spreading his promise of life through the valley, transforming the drab colours and bringing the dark bluffs into sharp contrast with the sparkling snow. A kea's call came from high up and echoed through the bluffs. The mountain day had really begun.

In the sun-warmed shelter of some rocks at the foot of the Bridal Veil Falls they stopped for smoko — the Falls at their back a mere trickle over black rock. In front, the snow-covered slopes ran up to the amphitheatre of bluffs below the Cone — down these, where normally many miniature creeks would be falling, hung festoons of icicles between towering pillars of ice. Arrayed like so many gigantic sculptured Grecian pillars supporting the upper plateau, they sparkled and glistened in the sun — from those bluffs catching the full force of the sun came the occasional tinkle of falling icicles.

As they rose to go, the head shepherd said to the boy, "Let's have a good bark up, you may see something it'll pay to remember." They let out a concerted yell and the dogs burst into full-throated support — the echoes bouncing back off the bluffs and ringing through the valley.

Suddenly there came a crackling from one of the bluffs, smaller icicles fell in dozens and the boy was amazed to see one of the great pillars of ice lean drunkenly out from the rock, to collapse with a roar in great chunks of ice and, sledging a mass of snow in front, go sliding down to the creek — more cracking and another great pillar joined its partner in the creek.

Then silence, broken only by the occasional tinkle of a falling icicle.

The lesson, "Never bark up or create a noise under ice-covered bluffs", was dramatically imprinted. It was instrumental in preserving his life on a later occasion.

The boy was given the top beat, with strict instructions to be careful in the chutes and to contact the top man from the other side of the last saddle.

When he reached the snowline it proved to be nicely firm, so he set off on a line that would bring him out at the saddle and still allow him to look down on the widening tussock below.

He wasted no time, well aware that his beat largely set the time taken for the day's muster. He saw no sheep until well down the face and they required little urging to go pelting off down a spur. Below and behind, he heard the others barking up. The day's muster was in full swing.

With the old dungaree coat off and in the sugarbag pack, he travelled fast, intent on doing a good job. He noticed the rising force of the wind and gave a passing thought to the cloud that had settled along the high top, a cloud that despite the wind did not appear to be moving.

The face bathed in the strong sun and the snow sparkling underfoot. Snowclad peaks stretching out to the west. The four dogs stringing along behind. It was a man's world with a man's job to be done.

When he reached the chutes he was careful to select the easiest place to cross, as much for the safety of the dogs as for himself. They were hard on the dark side and he took time to rip out steps with the horseshoe nails in his heels — a slip down into those blue depths did not bear thinking about.

Once clear of the chutes and still in brilliant sunshine, he angled up for the cloud-covered saddle where he had to contact the man from the other side — the higher he rose the stronger and colder grew the wind and the old coat was soon replaced. As he came below the saddle and rose up into the cloud, he was amazed to find that, in a matter of a few chains, he had passed from sunshine

Gentlemen of the high country. One of the world's hardest professions.

to a raging blizzard — the cold was intense as he pressed on for the saddle.

When he topped the saddle the wind was funnelling through with such force that it literally picked the dogs up and blew them over the top to send them sliding and scratching down the lee side. Leaning heavily on his hillstick he was blown, sliding, after them, down to the partial shelter of a rock.

Here he crouched, the snow hissing and swirling over and around the rock. It was the sound of the wind that hammered home the seriousness of his position — it thundered through the saddle in a great continuous vibrating roar, like the roll of a thousand great drums. The sound was so great it numbed his senses — the hissing, swirling clouds of fine snow cut at his face and eyes and made it difficult to breathe. Every instinct bade him seek shelter under the rock with the dogs, where they were curled up in a tight group, already almost completely covered with snow. But reason said no, you give in and you'll die there.

Brief glimpses he caught of the basin below, revealed a nightmare world of black rock and swirling snow. What if the other man was waiting there? He shouted, but could not even hear his own voice.

Wondering what best to do, he sought a drink of tea from the old whisky bottle in its sock in the bag. It was ice-cold, and every movement allowed the fine snow to drive in, up his sleeves and down his neck — he would have to abandon hope of contacting the other man, get back up over the saddle and let the head shepherd know the position.

He left the shelter of the rock and was immediately driven to his knees to avoid being blown down the hill. Scrambling on hands and knees up the ice-covered scree from which the wind had stripped the snow, he suddenly realised the dogs were not following. Backing down to the rock he could barely make them out under the snow — it was no time for half measures, if they were left there they might very well die. Shouting soundlessly, he waded into them with the hillstick, driving them out into the open, and turned to face the climb once more.

On hands and knees he went, clawing and scratching at the scree, blinded and barely able to breathe in the driving snow and ice. He paid no attention to torn nails or cut knees. One thought only filled his mind: up, over, and down. Past feeling, past caring, just keep on. Up, over, and down.

After what seemed an age he realised he was heading downhill in a lessening wind. Stumbling to his feet and leaning on the wind he set off blindly straight downhill. The wind lessened, the snow ceased, and he broke out of the cloud on to the sunlit face. All the dogs were there, shaking the snow from their coats.

Only then did he notice the hand in which he held the hillstick — it was dead white, with no feeling and he could not move it, the knuckles torn and grazed but not bleeding. Hell, it's frozen, he realised. Working the stick out through the clasping fingers with his other hand, he thrust the hand inside his shirt under his armpit and moved on down the hill.

Shortly, he got a reply to his barking-up from the head shepherd, waiting on a point lower down.

The head shepherd stared in amazement as the boy approached. His eyes were bloodshot and almost closed, his cheeks appeared to have been sandpapered and were flecked with blood, and he was almost doubled up with the pain in his hand.

After he heard the boy's story he gave him hot tea from his flask, checked his hand and sent him on round the face to meet the other men from the back with, best of all, the assurance that no disruption would accrue to the muster through his having failed to make contact at the saddle.

On round the face, drying out and warming up in the sun and wind, and with plenty of sheep work now to take his mind off his recent experience and the throbbing of his hand.

He compared notes with the man from the Back Face when they met and found they had only gone a short way up the back, because of the conditions, and had been nowhere near the saddle.

The mobs of lively, fast-moving wethers grew in size as they became concentrated between the closing arms of the Wye, soon to form into one large restless mob poised on the steep face above the crossing. Here considerable judgment and care were required to control their flow down to and across the creek — where the head shepherd stood, up to his knees in the rushing water, rescuing those unlucky enough to slip off the crossing boulders before they could be swept over the falls below. Then out and down the track in a long string to the yards below.

A memorable day for the boy. And one of which his hand never fails to remind him every frosty morning.

18

THE SHEARING

The gang is here,
The shed's alive,
It's "Wool-away"
As the shearers strive
To match the speed of the racing "gun".
There'll be record tallies when day is done.

CARRYING the scent of dew-damp tussock and the promise of a fair clear day, the soft airs of an early summer morning gently brushed the still waters of the lake along the length of curving white gravel beach.

The rays of the as yet unseen sun fanned out through gaps in the jagged skyline of the Remarkables, slicing golden paths through the blue mists of morning that softened the rugged western face.

An air of peace lay over valley and lake — but there would be little peace on the station this day, for it was the first day of shearing.

It was 5.30 a m as the Skipper and Bunky Henderson made their way towards the big old woolshed, after enjoying an early cup of tea with the shearers' cook.

As they slid aside the main door they were greeted with a flow of warm air, rich with the distinctively pleasant smell of healthy sheep, and a rattle of movement as 900 Merino ewes came alert to the sudden noise.

Shearing sheds possess a unique atmosphere, perhaps sensed only by sheep men. These sheds exist between two violently contrasting extremes — when not in use their cavernous and shadowy interiors stand empty and silent, except perhaps for the strident chirp of a lone intruding sparrow or the crinkle of corrugated iron shifting to a change of temperature.

Yet are they filled with the impressions of the relatively few days in the year when they are alive with the hustle and bustle of shearing or crutching, the shouts and chatter of a busy gang and the movement of thousands of sheep.

This old shed is no exception. With its first part having been built in 1863 literally hundreds of thousands of sheep have passed over its shearing board and hundreds of men of widely varying character have laboured at porthole, wooltable and woolpress.

Modern shearing methods. A good merino fleece is the Rolls Royce of wools. *Photo: K. & J. Bigwood.*

Originally, shearing boards had existed on both sides of the shed with eight stands a side. Somewhere about 1915 the decision had been taken to introduce the then revolutionary concept of machine shearing and an eight-stand shearing plant was installed on one of the boards. This plant had been none other than the world-famous Bergan & Ball model. It must have been a headache to maintain with its white-metal oil-fed bearings and the long unbalanced downarms, awkward to work with.

Legend has it that the owner looked forward to big tallies and short shearings through introducing a competitive element into operations by having one board use machines and the other the blades. Somehow, human nature being what it is, matters did not work out just as expected.

On one side of the shed the blade board carried on in the well-tried and trusted style in conditions of reasonable peace; from across the intervening catching pens and races came the incessant, nerve-stretching rattle and clatter of the machines, that portion of the shed being all too often shrouded in a blue haze of smoke from overheating bearings.

Breakdowns and teething problems with the newfangled invention invited ribald remarks from the blade board. Instead of friendly rivalry, bitter antagonism developed between the exponents of the differing forms of the art, the occasional scraps threatening to erupt into a major conflict.

When shearing eventually ground through to a finish the owner had had more than enough of this type of competitive stimulus. Pinning his faith on the modern approach and no doubt influenced by the capital expenditure involved in establishing the machines, he nailed down the portholes on the blade side and replaced the shearing board with sheep pens.

However his faith in the new process was shortlived, for another aspect of machine shearing was soon to be brought home to him. In 1917 a disastrous snowstorm struck during shearing and the closely shorn sheep died in thousands.

The machines were closed down, never to be used again, and operations reverted to an eight-man blade gang until 1956 when a modern four-stand machine plant was installed.

Now the Skipper and Bunky walked along the top rails between the pens of moderately packed sheep in the shed. They sought for any sheep that might have been lying down and got trapped under the others as they surged to their feet, for the pressure created by the small neat feet of the Merino can be likened to that of the stiletto heel of a woman's shoe, and no sheep would last long under these trampling hooves once the bustle of the day's shearing got underway.

There were still a few odd jobs to attend to before the Skipper could be satisfied that all would go smoothly. Branding paint had to be mixed, paint pots, brands, raddle and yard books made ready at the branding-race shed. The fastenings on counting-pens and holding yards checked — all part of the thousand and one jobs that have to be attended to to ensure a smooth operation.

Suddenly the big Burmese gong at the cookshop boomed out, announcing to all and sundry that breakfast was ready and the day had begun.

The Skipper and Bunky made their way back through the shed, the Skipper pausing on the shearing board to cast a final glance around the central scene of operations. The long shearing board, the wooltable, woolbins and woolroom floor lay unsullied, clean and white from yesterday's scrub down; the eight water pots filled with disinfectant, and wool to protect the shears were ready beside each porthole. Tar in the tar pots, oilstone pads of close-clipped sheepskin, pieces-baskets, tally and wool books, all were ready. Satisfied, they moved on to breakfast.

The old cookshop and quarters had tons of character. Built to an L-shaped plan, the long side of the L formed the diningroom with the cook's quarters and the kitchen, dominated by a huge freestanding coal range, at the top end. This was the unquestioned domain of the cook, the autocrat of the shearing gang. Woe betide any green rousy who committed trespass; even the old hands treated this area with due respect. A wide serving hatch ensured that the kitchen remained sacrosanct.

Almost the full length of the fifty-foot-long diningroom was taken up by a linoleum-covered table flanked by wooden forms. At the bottom end were two doors, one leading through to the shearers' bunkhouse and the other outside. Here also was located a six-foot-wide fireplace and a colonial oven, relics of the original building, which now did duty as a diningroom. This room and the central part of the woolshed had been built in the late 1860s and were still roofed by hand-split red beech shingles, though now covered by the more efficient but less picturesque corrugated iron.

A strict order of seating existed by common consent. The Skipper sat next the kitchen on one side of the table; then, in descending order, the head shepherd, the penner-up, the brander, and any shepherds present. The other side was headed by the classer, then the wool-roller, the shearers' rep. and the older shearers. Below these on both sides sat the other shearers, the pressers followed by the rouseabouts. Below these, the lowest in order of importance, sat the 10-year-old boy, for he had the least prestigious job of all, clipping dags, running messages and milking the cow. But he would not have swapped places with a king; he was part of the whole exciting operation and receiving the princely wage of twelve bob a week.

This first full meal established the seating at table for the remainder of the shearing, casuals arriving for a meal thereafter had to take their place with the boy at the bottom end.

As men finished their meal they wasted no time in moving over to the woolshed, for today was the beginning of a major run of shearing. Between 55,000 and 60,000 sheep, on five separate stations, lay ahead to be shorn. This was recognised as a good run among the shearers and applications for a pen (the right to shear) had closed months ago.

The shearers made straight for the stands they had selected the day before, to run their oilstones just once more over blades (shears) already more than razor-sharp.

They had received their two free pairs of blades and a bottle of salad oil from the Skipper the previous day and had spent many hours at the two big grindstones "taking them down", attaching drivers, jockeys, spurs and knockers. Pulling them back to give a wider blow, adjusting both spring and set until they had a perfect if deadly tool just right for their individual requirements.

They were all dressed in the recognised style, blue sleeveless

The *Earnslaw, Ben Lomond* and the *Mountaineer* provided a very good service on the lake. The *Earnslaw* taking on a load of wool at the Woolshed.

bush singlets, saddle-tweed or white moleskin trousers supported by bowyangs below the knee, and moccasins made from woolbale caps, the latter a costly perk to the runholder, but one established by long custom.

Shearing, especially blade shearing, is hard slogging toil made easier only through perfection of style, and is a highly specialised trade. The shearer is paid by the 100 sheep shorn, so the greater the speed he can acquire the more he will earn during the season. Yet his own pride and the standard set by the runholder will require that the sheep be turned out well: smoothly and evenly shorn, with no nicks or cuts.

His interest in the job, apart from the considerable amounts that can be earned, is maintained by the fact that no two sheep are exactly alike. Some are flyers and others sticky, and the perfection of his shearing technique must cope with the wide variety encountered.

The shearer also has his professional pride in being the "ringer": the man who has shorn the most sheep from any particular shed is said to have "rung" it, as is the man with the highest tally for a particular day.

Characters are as diverse in a shearing gang as in any other group of men and this one was no exception. Yesterday a tall rangy chap in the uniform of a Salvation Army officer had walked into the shearers' bunkhouse and dropped his bag on a bunk. A slightly uneasy silence settled over the room, as the newcomer proceeded to unpack and change, until a hard case could stand it no longer and asked "What the hell are *you* doing here?"

He was told quietly that money was needed for a hostel and he thought this a quick way to raise some so he'd taken a pen for the run. This was greeted with a stunned silence which was broken by laughter and some clever cracks when the newcomer went out to report to the Skipper.

Some reckoned the Skipper must be getting soft in giving this bloke a pen, but one of the old hands said, "Better be careful, he may be able to shear. Last year some of the lads bet a publican in Queenstown he couldn't shear a sheep in under ten minutes. He was just a little thin bloke, and at a pound a head there were plenty of takers. Well, when they asked the Skipper if it would be OK to have him out, he'd said with a smile 'Yes, tell him to be here for afternoon smoko,' and no, thanks all the same, he didn't want to make an easy quid.

"Well, this dapper little bloke turned up at smoko and all hands gathered to watch the fun. He asked if one of the lads would shear a sheep first so he could see how it was done and give him a bit of instruction. Tom here obliged — didn't you Tom? — and showed him which hand to hold the blades in as well. Then he asked you to carry the ewe out for him in case it proved too much for him. A ewe! And the shed was packed with those big wethers they've got here — you wait till you see them, they're so big they've got to get down on their knees to get out of the porthole. Well, Tom brings out the biggest and roughest wether in the pen and hands it to the little bloke. When he tries to lift it up to sit it down it's taller than he is and it gets away from him, and the lads have to catch it and bring it back. This time he gets it on its side and sits down on top of it to hold it down. Then he says he's still willing to try, and would the classer be kind enough to ring the bell to start him off like a real shearer. But 'Hold on a minute,' says Tom, 'how about doubling the bets?' 'Well,' says the little bloke, 'that's making it a bit hot, but I'll oblige you.' And the others took him up too. Then he slips off his coat and hands it to a rousy and nods to the classer.

"Bang goes the bell and the little bloke whips the wether upright and the blades seemed to jump out of the pot into his hand. He had the belly off before the lads had drawn more than a couple of breaths, and that wether went out of the porthole in four minutes thirty flat — and nicely shorn too.

"There was dead silence on the board as the little bloke slid the blades back into the pot and said, 'If you care to come to the hotel on Saturday night you can settle up then and I'll shout for you,' and walks off calm as you like to have a cup of tea with the Skipper. It turned out he'd been a real gun who'd shorn for the Skipper years ago. You should have seen the lads' faces when it sank in just how many of those big wethers they had to shear to square that lot."

Now all hands gathered to watch the start of shearing; the Skipper talking quietly with the classer, the shed hands and rousy standing about arguing amongst themselves as to who would be the fastest shearer and how the "Salvo" wouldn't make enough to buy tobacco. Old Bunky sitting on the top rail of the race, keeping an eagle eye on the tightly packed catching pens and sending up clouds of smoke from his short black pipe.

The shearers leant over pen-doors trying to sight an easy one for a starter.

As the minute hand of the big clock crept closer to starting time and the classer picked up the donger to strike the starting bell, an electric silence settled over the shed.

Down the long board some shearers stood with a hand on the pen-door, the recognised guns (very fast shearers) casually inspected their fingernails or wound a ring of wool round a finger subject to blisters. The Salvation Army man on number 3 stand stood quietly with folded arms, and away down at the far end the learner had just discovered he hadn't tied his driver down and with trembling fingers was attempting to set it right.

CLANG went the bell, *BANG* went the pen-doors, and in a flash eight sheep were clamped between knees as hands dived for the blades. With a silken hiss the blades bit into the belly wool and first one then other bellies were tossed aside to be grabbed by the picker-ups and flung into the belly bin. Gosh, look at number 3 go, five blows to the belly and he's already going up the back, his shears taking great gulping blows and the fleece fairly falling off the sheep. Another minute and it goes bounding out the porthole, beautifully shorn. First off the board in a bit over three and a half. There goes number 1's, and others follow in quick succession. There are cries of "Wool away" and "Broom", the rousies race to clear the board and fling the fleeces high and wide upon the table, where the practised fingers of the rollers quickly strip away the neck and pieces, roll the fleece and hand it to the classer, who is now absorbed in setting out his lines.

This looks like being a fast gang and the Skipper's thoughts dwell fondly on the possibility of better than twelve hundred head a day and a short shearing.

As number 1 is casually dragging out his fourth sheep, the learner's first sheep gets away from him and comes rushing up the board dragging miles of wool, pursued by its redfaced barber. The Skipper turns his head away and appears not to notice as the discomfited learner drags his quarry away to finish the operation. The older shearers chuckle and pass a few unhelpful remarks. The electric tension of the start is broken, the pace steadies down, and the shed settles into the smooth rhythm of experts busy at their job.

19

GENTLE GIANTS

The example set
Is seen not now,
Of horse and man
With disc and plough,
Tilling the good dark acres' soil
In harmony of mutual toil.

IT was half past five on a fine frosty morning in the early fall — high above, the peaks were edged in gold from the rising sun, but here in the valley a light frost whitened the grass and a wisp or two of mist hung in the hollows.

The sheep had not yet risen from their camps, they rested in small groups, each one with the lightly frosted mark on the ground beside it, clear evidence of its having risen halfway through the night to turn round, as is their practice. The valley was still at rest, reluctant to start the day on this crisp morning.

All hands in the three rooms of the hut were fast in the deep sleep that precedes the dawn, when there came a thunderous knock on the door of the teamsters' quarters. Again it came, more urgent, the door shaking under the force. The teamster, struggling back to consciousness, sat up in his bunk and peered at the clock in the dim light. Again the knock came, more prolonged, and he roared out, "Get the hell out of here and leave a man alone." There came the sound of departing footsteps and he snuggled back under the warm blankets.

Ten minutes passed and he had just drifted back to sleep when the knock came again. With a curse he threw back the blankets and, with shirttails flying, rushed to the door, flinging it open. "I'll wrap a standard round your rump, you cheeky little bastard," he bellowed in simulated anger. Outside, with ears pricked forward and an innocent expression on his face stood Punch, the nuggety little draught horse. Beside him towered Baldy, the big shafter — they turned and, with the complacent air of a duty well done, clumped off over the ridge to join their fellows waiting at the stable to be fed.

As the teamster began dressing, there came the muffled clatter of an alarm clock, abruptly stifled to the accompaniment of a muttered curse. Next door, the boys turned over in their warm

Gentle giants of yesterday. "The power and stamina of a good team were considerable by any standards." Wool leaving Glencoe for Frankton wharf. J. Stevenson, driver.

bunks with sleepy grins, for, as Punch's pawing at the door always beat the alarm by a matter of minutes, they derived a great deal of innocent pleasure from this almost daily pantomime.

The old hut stands empty now, a broken window here and there and starlings busy in the chimney — but the marks of Punch's early morning visits are scored deep in the sagging door, mute evidence of the part the draught horse played during that period of the station's history.

Two six- and one four-horse teams, plus replacements, formed the complement of this working force. The Skipper was an enthusiastic and expert horseman, all the draughts and hacks being bred on the property. Teams of draughts and a string of hacks and packhorses were always entered in the local shows — the rows of faded prize cards and now tattered championship ribbons still hanging in the old stable bear silent witness to his judgment and the expertise of the men who worked with and tended the horses.

The teamster's day began with the before-breakfast feeding and grooming of his charges in the warm intimacy of the long, cobble-floored stable, to the accompaniment of contented chomping of good oaten chaff. By the time the teamster had finished his own breakfast, returned and slipped on the collar-and-hames and winkers, they would be finished eating, ready for the day's work in plough or waggon.

The good teamster had a close rapport with these (mainly) amiable giants — they exhibited a wide range of character, ability, likes and dislikes. There was a considerable art in selecting and positioning a team, in the creation of a contented, efficient unit of smooth flexible power — and the power and stamina of a good team were considerable by any standards.

When the first crawler tractor arrived on the station it was looked over and greeted with scorn by the teamster, as befitted a proper horseman. The blacksmith, now promoted to part-time mechanic/driver, polished the shining bonnet and made various forecasts on the superiority of the new contraption, all hotly denied and argued over by the teamster. Rivalry and friction between the two continued to increase, reaching a point where it looked as if the question would be settled by other than academic argument.

At this stage the Skipper, who had endeavoured to maintain some sort of peace by having them work widely separated paddocks, decided there was only one way to save the day and settle the argument beyond doubt. He called the two together one lunch hour and, as diplomatically as he could, indicated that he wished to have a certain 70-acre paddock ploughed as quickly as possible for the sowing of turnips and that they were to work it together. Also, as he was interested in testing the capabilities of the new tractor, they were to plough half each and were to start from opposite sides of the field, working exactly the same hours. This would allow the horses to become gradually accustomed to the tractor as the work progressed, for they had, if possible, a more violent dislike of the clattering, squealing and banging of the crawler than had their master.

The participants had also to take their ploughs and what other gear they required down to the paddock that afternoon, peg out their first land and start work at 7.30 next morning, knocking off at 5 with the appropriate breaks for smoko and lunch in between.

He would himself come down to the paddock later in the afternoon to measure and peg out the halfway mark. As the ploughs were both three-furrow jobs of the same make and the paddock was flat, it would be an even competition: the competitors would be the pride of Detroit, USA versus the honour of the horse.

When the news of the arrangement filtered through to the boys via the cook at tea that night, the arguments blossomed forth on a wider base, some favouring the tractor and some the horses. It was almost as interesting as the Melbourne Cup and some heavy bets of tobacco were taken up.

The next morning Punch had no need to knock up his master. The horses were mildly surprised to be invited in for their feed before the stars had left the sky — and they were given double rations to boot.

The first day passed without incident, one of the boys reporting back that evening that the horses were one round in front.

At lunch the next day the report was from one who had, by chance of course, ridden back that way: the tractor had now caught

"The beauty of dignity, the essential truth, of man and animal toiling in harmony together with nature."

up and was a round and a half in front. That evening a committee inspected the "track" and found the tractor was now four rounds in front — the driver had accomplished this by slamming into a higher gear on the slightest downhill grade. The head shepherd reported this had caused the tractor to give out great clouds of smoke from the exhaust. Seeing the smoke rising, he had thought a fire had broken out somewhere and had cantered over to check, only to find it was due to the aforementioned phenomenon.

The teamster was up bright and early the next morning and the boys were intrigued to see him leave with twelve horses — he had taken a spare team and would be swapping over at lunchtime. The tractor driver was muttering darkly about unfair practices, but was informed in no uncertain terms that as he changed the oil in his tractor, so the teamster had every right to a change of horses.

The latest bulletin at lunchtime was that the horses were going great guns and had already caught up with and passed the tractor which, it had been noted, had stopped for a period with the driver tinkering with his plough.

That afternoon two of the boys could not resist visiting the scene of battle. The tractor was grinding and clanking along, issuing forth great clouds of smoke under the strain of pulling a plough which had had its furrows extended to maximum width; the horses were skipping along, apparently with ease, closer inspection revealing that the plough depth had been reduced from eight to four inches.

As they were watching the action they spotted an approaching rider and, recognising the Skipper, decided they had urgent business elsewhere. They reported that, as seen from the top of a distant rise, there appeared to be a conference taking place in the centre of the paddock. An additional late report indicated that the ploughing had suddenly reverted to normal practice.

Late the following afternoon a courier was despatched to check on the situation, with strict instructions to reconnoitre the area first in case there was an official present. He reported later all had been clear and that he had arrived at the critical moment — one plough-width of a split finish remained with the centre markers right up the middle, first competitor to drop his plough into this would be the winner — the tractor was rattling flat out across the bottom end of a land and the horses were just lifting out at the end of theirs. The tractor looked to be an easy winner. Then suddenly there came a whistle and yell from the teamster, the six-horse team broke into a trot, and with a great old clinking of chains and in a cloud of dust, swept round the land to be pulled up with a jerk as the plough was dropped in to complete the finish.

When appealed to the Skipper adjudicated it to be a draw. Also, for the record, the experiment failed to put an end to the argument. . . .

Baldy and Punch were mates, inseparable out of harness and only truly content during work when the other was somewhere in the team.

A more ill-matched pair would be difficult to imagine — Baldy, a truly huge horse, wall-eyed to the point of ugliness, deliberate in his movements like an elephant, a quieter and more gentle horse there could not be, and a wonderful shafter.

Punch, on the other hand, was small and nuggety, with a heart like a lion's. His chains would be as tight at the end of the day as they were at the start. Out of harness he was always on the move, up to some mischief or another, or simply pestering old Baldy.

Baldy in the shafts was a joy to watch, his chains always looked tight but a baby could have put its finger between the links without getting hurt, that is, until Baldy was called on to pull, when he would settle into the chains with a grunt and almost straighten the links right out.

One day all hands were carting chaff to the loft. Baldy was shafter in the big dray, with two other horses in chains in front. The dray, parked below the loft, was piled high above the frame with heavy bags of chaff which were being tossed up by two men — the driver had omitted to tie the leaders back with the reins and suddenly, for some unknown reason, they took off. The two men went off the back of the load with some bags, the corner of the dray frame caught in the barn wall — ripping down a section with

The author's father bred all his own draughts and hacks, which competed in local shows. "The rows of faded prize cards and tattered ribbons hanging in the old stable bear silent witness to his judgement."

a great splintering crash — and they were away, flat to the boards, bags of chaff flying everywhere. Up the hill they flew, with Baldy a very surprised-looking participant. At the top the leaders wheeled away from a steep gully and started back down, gathering speed at every jump. And this is when Baldy evidently decided this foolishness had gone far enough — he simply sat back in the breeching and in less than three chains he had hauled those panicked horses to a halt and held them there, still lunging and snapping at the chains, until someone ran in and gathered up the reins. That sort of violent exertion was not for Baldy.

The teamster, through his team, had a close affinity with both soil and Nature and, perhaps through this, derived a greater satisfaction from his work than is possible for the tractor driver of today. But to have hinted at this to a man smothered in dust and streaked with sweat, from eight hours following harrows behind a fast-stepping team, would have invited hoots of derisive agreement.

With mechanisation man has gained greater speed, ease and possibly efficiency in his agricultural pursuits. But there have been losses too — indefinable losses by man in general, losses that can be likened to those incurred by the passing of the graceful clipper ships from off the seven seas.

One is saddened by the thought that, with the disappearance of these noble teams from the fields, youth will be denied yet one further opportunity of witnessing the beauty and dignity, the essential truth of man and animal toiling in harmony; together with Nature.

20

"I TELL YOU . . ."

OLD BOB was nightwatchman for a big mercantile company in the city — invariably dressed in heavy, grey, baggy-kneed, saddle-tweed pants, a grey shirt with well-used black tie beneath an ancient grey waistcoat. Tall and thin as a reed, he walked with a slight stoop, his long calm aquiline face below the shock of white hair slightly bowed as if in memory of many encounters with low beams during the dark of his rounds. He has long since passed on, but the memory of his quiet and kindly nature still lingers with those who knew him.

It is difficult to visualise what would have happened had he been called upon to face intruders, for a more gentle person would be hard to find. Perhaps the sight of his shadow, the big Alsatian, would have scared them off; but this too is doubtful for, through long contact, the dog appeared to have become completely identified with Old Bob's nature. He was a huge dog, if slightly overfat, who padded after his master on their brief daylight appearances in the late afternoon, slowly wagging his tail, his big wet tongue dispensing good fellowship to all and sundry, strangers or not. The store and woolroom cats were also part of the family group, two or three of them, tails erect and loudly purring, usually forming a satellite escort, rubbing themselves indiscriminately against and between the dog's and Bob's legs.

Bob's home was little better than a cubbyhole in the corner of the big woolroom. To reach it one had to negotiate a narrow canyonlike passageway between the high stacks of woolbales which so completely surrounded it that it was little better than a windowless cave among the bales. Lit by one glaring electric bulb, at one end was Bob's neatly made bunk, with, at its feet, a big flat packing case filled with old sacks — this was the dog's bed and also the refuge of any of the cats that might be feeling the cold.

A small table with a couple of chairs and a single-bar heater close by filled most of the centre space; the other end was taken up by a long bench which bore a bucket of water and a basin, a single-element cooker, an assortment of full and partly filled milk bottles, a couple of plates automatically drying as they leant against an upturned mug. The bare board walls were thickly covered with illustrations cut from the *Otago Witness* and newspaper cuttings, mostly relating to the 1914-18 war, but interspersed

While a boy, the author had to spend several months in a Dunedin hospital. The enforced idleness and the kindness of an old friend endowed him with a lifelong love of reading.

with articles dealing with practically every subject under the sun. This was Bob's home, but it would be not entirely correct to say it was where he lived, for every available spare space was stacked with books. There were books under the bunk, on the table and bench and tiered along the walls, books on every subject imaginable and of all types, bound gilt-edged volumes that must have cost a pound or two were stacked indiscriminately with dog-eared paperbacks and periodicals, novels, books on travel and the sciences. All were grist to the mill of Bob's enquiring mind. From this stuffy little room, filled with the not unpleasant smell of jute and wool, to which the sounds of the busy outside world came only faintly through the insulating piles of bales, Bob's mind could range at will throughout the world through the medium of his books. Apart from a few shillings spent on food, all his small wage went in books: they were his passion and his life.

The Skipper had known Old Bob for many years and a mutual respect had grown up between the two, so different in character and mode of life. On his infrequent business trips to the city the Skipper always made a point of going down to the woolstore for a few words with Bob. Once or twice the boy was in tow, taken to

"As surely as Winter falls, so will come the Spring."

the city to learn something of the outside world and to give his mother some respite. Bob had taken a liking to the boy and a parcel of books was always quickly made up to be pressed into the hands of the Skipper and duly delivered on his return.

It was about this time the boy went down with rheumatic fever, ending up confined to bed for several months in a hospital in Dunedin. Bob came to see him twice a week, with a bag of sweets and a bundle of books. So many were delivered that the nurses had to take a firm stand and demand their removal.

After his release from hospital the boy was allowed home but was forced to spend the greatest part of a year almost entirely lying down resting. Every month a kerosene-case full of books would arrive by rail from Old Bob. With little else to occupy his time it is little wonder the boy soon amassed a wide general knowledge and an enquiring mind, plus quite a library of widely assorted books.

In this collection was a book called *The Child's Bible,* given him by his mother — designed for children, it still contained the basic principles of Christian faith and, to the boy, made pretty dry and unlikely reading. However, after many samplings he eventually read it right through. A lot of it seemed more unlikely than some of the novels he had read, and he could not fathom out why, if what the book said was correct, religion did not play a much greater part in everyday life. Why did people clam up, like they did when you mentioned marriage and things?

To his very young but critical mind, if what was said in the book was correct, then there must be some tangible way in which it could be proved. One of the passages had stuck in his mind: "Whatsoever you shall ask in My name, shall be given you." Here, decided the boy in his innocence, was the perfect way to obtain confirmation, one way or the other — but it must all be done in the proper form and not leave room for any future doubt. So, after due deliberation on procedure, he hopped out of bed, knelt down and, clasping his hands in prayer, asked God in the name of Jesus to send him a sign so that he might believe.

Unaware of the enormity of his request he climbed back into bed and blew out the light, satisfied in mind that he had taken the only proper course to sort the matter out. He drifted off into a dreamless sleep, lulled by the muffled rumble of the falls below the house.

Later that night the boy came gradually awake. Lying there in the darkness he became aware that the room seemed to be filling with an overwhelming, living sense of power — a friendly, loving, compassionate power that flowed into and through him.

In the void of darkness above him a faint illumination grew in strength and nearness, to become, quite clearly seen, a person's head looking down at him. Long wavy hair fell to the barely seen shoulders, and the lightly bearded face, filled with an ineffable compassionate kindness, smiled down at him.

The boy lay there, completely unafraid, gazing up at the face, which almost immediately began to fade — as it faded, so too did that flowing sense of power, till all was as it had ever been.

Only the darkness and the familiar rumble of the falls. But the boy was left with a great feeling of peace and certainty, to lie there in wonderment until he drifted back to sleep.

The years have rolled by and the peace has often been lost, but never the certainty. The man who was once a boy knows full well that always after the summer comes the winter: how long, or how hard, none can tell. But as surely as Winter falls, so will come the Spring.

God be with you, go in peace.

PART TWO

THE SAGA

In tribute to those who first braved the unknown seas and to the many unsung pioneers of these our mountain lands.

What if I live no more those kingly days,
Their night sleeps with me still.
I dream my feet upon the starry ways;
My heart rests in the hill.
I may not grudge the little left undone,
I hold the heights, I keep the dreams I won.

Geoffrey Wintrop Young
Mountaineer

THE SAGA

It is enough we sight the grove, then shall we find the tree we seek. POLYNESIAN NAVIGATORS' PROVERB

i. Children of the Mists

KUPE, that great Pacific navigator and voyager of long ago, had followed the recommended star-path south over a thousand leagues of rolling sea, from his island home where the people were now thick in the valleys to this southern area, where the legends of other exploratory voyages recorded the existence of large islands.

Yesterday the wave patterns, the life of sea and air, had told him they were close to land and he had altered course to follow the evening flight of homing birds. Now it was morning and the two great voyaging canoes, urged on by a gentle breeze, were just whispering along over the long ocean swells.

As the light strengthened and the sea mists lifted, Kupe saw, far to the west, what appeared to be a long bank of white cloud.

As the morning passed and the canoes gathered speed before the freshening wind, it became apparent that this was no cloudbank, but land reaching higher than any he had known — a land white on top, with brown hills and green valleys sweeping down to the sea. He named it Aotearoa — Land of the Long White Cloud.

There can be no certainty as to who were the first people to tread the shining sands of our numberless beaches and to call this land their own.

Man has ever sought to reach forth beyond the horizon, be it mountain pass or rolling ocean swell; or to the dim galaxies beckoning from beyond the stars in the boundless depths of space.

Before even the Phoenicians were carving out their trade routes in the boisterous waters of the Mediterranean and the South Atlantic, before the Vikings came storming down out of the grey mists of the north, the people of the Pacific basin were reaching out across immeasurable leagues to seek and settle island after island.

Over the great ocean roads of the Pacific spread the restless voyagers of long ago — the Polynesians and, to a lesser degree, the Melanesians, were a people of the sea, at home on the heaving bosom of Hine-moana, the Ocean Maid.

With their *mauri*, their life-force, purified and concentrated in some tangible object; secure in the knowledge they were under the protection of Kiwa, one of the guardians of the Ocean Maid, and confident in the ability of their navigator, they set forth on voyages whose full extent has yet to be discovered.

These navigators were specialists. They were the repository of generations of an accumulated knowledge of navigation by means of astral and natural observation. Their methods and ability have been proved beyond doubt; they were not "pinpoint" navigators, for with their ability to read natural signs it was sufficient for them to reach the immediate area of their desired landfall to be certain of the successful completion of the voyage.

Particular voyages were commenced from recognised departure points, which provided leading-marks for reference until such time as the navigator was well and truly seaborne, clear of all interference from the land. From there on, the heavenly bodies, the wave and wind pattern and the behaviour of sea life indicated not only direction but latitude as well. Ever conscious of the old proverb "On sandy paths are footsteps seen, but who shall scan footprints on the far-flung ways of the Ocean Maid?", they relied not only on their own observations but on the vast store of knowledge passed down generation by generation — up to thirty-two points of the compass bore names, which were also applied to the winds from those directions; hundreds of stars bore individual names with their ascension and declination being memorised.

The great wave of exploratory expansion in the Pacific appears to have taken place within the first 1200 years AD, this being followed by a period of consolidation, with migrant voyages taking place over recognised routes between the then known lands.

The number of planned voyages to and from this land, to the islands in the north, cannot be stated with any degree of certainty, but that quite a number did take place has been established beyond reasonable doubt, by comparative investigation of legend and method. Added to these organised voyages must be the accidental arrival of refugees driven forth from their home islands by attackers or swept beyond reach of their shores by adverse winds.

The first of the organised voyages found people already here, the *tangata whenua,* people of the land — it can only be assumed that these people evolved from a succession of accidental arrivals over a long period, combining with a people whose presence more than 1500 years ago has only recently been hinted at by archaeological discoveries.

The *tangata whenua,* or the Moriori as they came to be known, were apparently a gentle and simple folk; they were easily overcome, enslaved and absorbed, in more ways than one, by the more aggressive newcomers. The final act of almost complete annihilation of the one remaining distinct group, resident in the Chatham Islands, occurred in 1835.

The trading brig *Rodney*, in search of whalebone and other trade goods, made Port Nicholson (Wellington), on 26 October 1835.

A very large number of Maoris were about the harbour at the time, apparently fearful of attack from tribes to the north and the master of the *Rodney* was coerced or conned (more likely the latter, for another vessel, the *Caroline,* was also in the harbour at the time) into transporting approximately 500 Maoris in large canoes to the Chathams — the Moriori offered no resistance and were quickly enslaved.

Five years later only a handful of Moriori remained out of a population previously estimated to exceed 1500. It is reported that this rapid decline was due more to the effect of an epidemic which reached the islands than to the practice of the Maori to eat any who opposed him. The last person considered to be of pure Moriori descent died about 1930.

At the time when the European settlement of New Zealand began, the admixture of Polynesian people already here lived under a strict tribal system and in an almost constant state of inter-tribal warfare. No generic term for these people appears to have evolved, they were known to the first European visitors as members of a particular tribe, or as "natives" or as "New Zealanders". With the passage of time the term Maori was adopted by and used in reference to these people.

The European settler, like his earlier counterpart, came from many a race and creed. Both had the same motivation: to seek escape from their troubled and overpopulated homelands and the chance to build a new life in a new land.

Emigration to this small country continues almost unabated to this day — it is against the historical background of settlement of this country that the following should be considered.

Firstly, unless the increase in population, whether permanent or transient, is strictly controlled, we are in danger of creating the very conditions our forefathers sought to escape — of losing a way of life that is the envy of the world and which thousands of this country's sons gave their lives to protect.

Secondly, there existed no original aboriginal or clearly defined ethnic group — no matter from what race we or our forefathers came, we are *all* New Zealanders. Any citizen of this country who seeks to be known, or allows himself to be referred to, as other than a New Zealander is undermining the foundations of a concept of human equality which is an example to all mankind.

We are New Zealanders — the *tangata whenua.*

The thrill these early voyagers received at this first sighting of of their promised land could not greatly exceed that experienced by the returning New Zealander, jetting homeward across the stormy Tasman Sea — for these shining peaks above the clouds are still the first of the land to hold forth its welcome and its challenge as of old.

As in the past, so today, do these islands draw people to seek, like Kupe, a new life in a new land. Widely varied as to origin, the people cherish an ideal of a free, understanding and integrated society.

In comparison with the increasing populations of older lands, these islands have remained relatively thinly populated and, for the sake of the quality of life the people seek, long may they remain so. The natural barriers that served to restrict the human tide are now no more, so man-made laws must maintain a balance, lest what is sought is for ever lost. For, unlike Kupe or our forbears, there are left no new lands to which our children now can turn.

ii. The Land

This group of islands that makes up the land of New Zealand is surely among the most favoured in the world. Small, seagirt, embracing contrasts of terrain and climate almost unbelievable in

"These shining peaks above the clouds", Aorangi, the monarch of Aotearoa.

so small an area, yet is it ideally suited to all aspects of the farming industry for which it has gained worldwide acclaim.

Down the western side of the South Island stands the great chain of the Southern Alps, their towering massifs clothed in eternal snows. West of the Main Divide the heavily bushclad hills fall abruptly to meet the Tasman Sea, leaving only restricted areas suitable for pastoral use. On the drier eastern side, the thousands of square miles of tussock-clad mountain lands, the fertile plains and valleys, are ideally suited to the now firmly established and expanding range of pastoral activities.

It is to these mountain lands buttressing the main alpine chain — in some cases encompassing it — and, more particularly, to that area within the Lake Wakatipu basin, in the Southern Lakes area of Otago, that the foregoing reminiscences of high-country station days have applied.

This great Wakatipu basin, with the mountains cradling the blue water of the lake, is undoubtedly one of the scenic gems of the world.

In the not too distant past, in the Ice Age, it was the point at which a huge valley glacier diverged, to send its major branch moving relentlessly out and down to the lower southern lands.

The last sudden and dramatic climatic change resulted in the rapid thawing of the icecaps and glaciers, leaving an enormous volume of water lying in the valley, dammed back by the terminal moraine at Kingston. Although the freshly formed lake originally had two simultaneous outlets, it eventually cut its present and only visible outlet through the Kawarau Gorge, to mingle with the other waters now rushing down from the thawing glaciers of the Wanaka Basin and jointly to form the mighty Clutha River.

With the deepening of the Kawarau Gorge outlet the lake retreated from much of the area covered and stabilised in its existing, roughly S-shaped form. With its surface at an altitude of 1013 feet above mean sea level and depths exceeding 1200 feet, much of the bottom lies below sea level. A peculiarity of the lake is its unexplained, regular, rise and fall of approximately three inches every five minutes.

The mountain scenery owes it existence, as does the whole of the Southern Alps, to the unbelievable pressures exerted by plate movement along the fault line that runs the length of the Alps and passes through the district. The isoclinally folded Tertiary rocks and spectacular formations so well displayed in the Remarkable mountain range provide a good example of the forces so freely wielded by Nature in the past.

A wild land this, in many places bordering on the perpetual snowline — towering peaks and plunging bluffs, swift-flowing icy

"A wild land this . . ." Below Lake Hope.

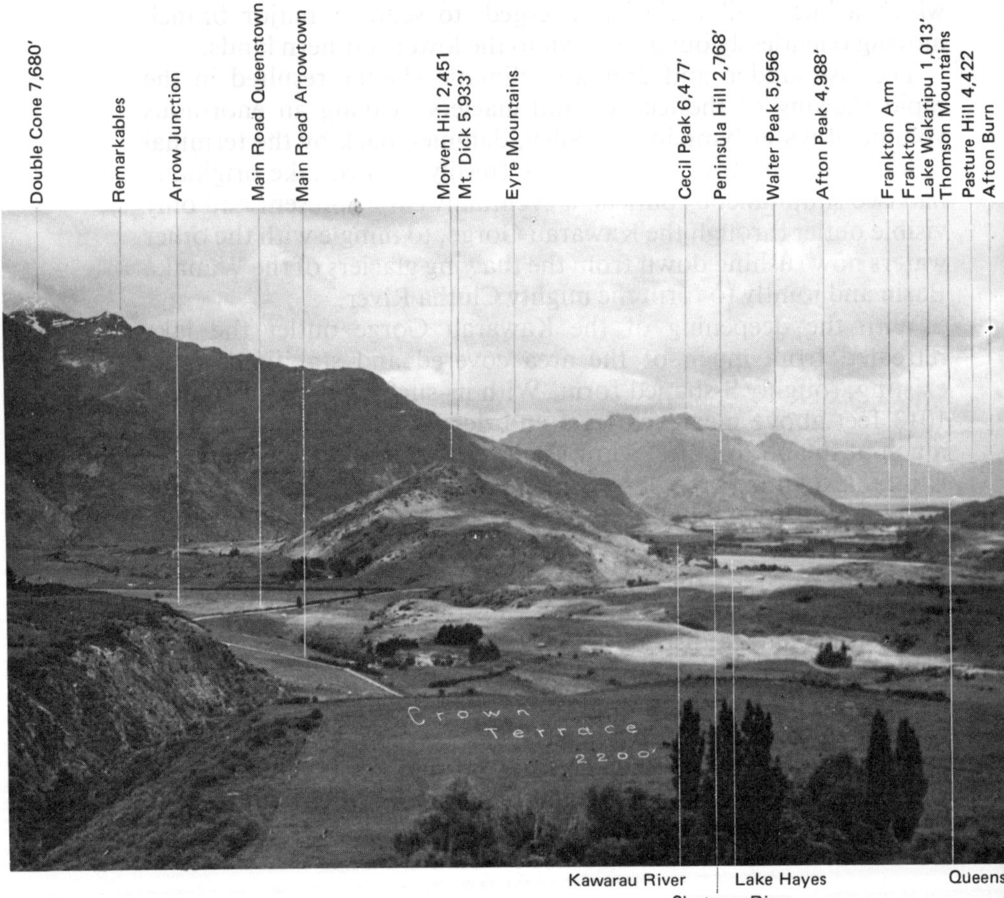

"This great Wakatipu basin" — The eastern sector of the 2,000 square mile catchment.

rivers and untold square miles of tussock, "gold and silver in the changing light"; the natural home of the kea and the falcon, where Nature wears no veil.

We can assume the mountains have remained, almost unchanged, in the passage of a thousand years. Not so the now-fertile farmlands and lower valleys which, a little over a century ago, were clothed in impenetrable thickets of matagouri scrub and dense fields of speargrass, the *taramea* of the Maori. Only those who have had no other choice but to force their pain-filled way through either of these can fully appreciate the daunting conditions that lay in the path of the travelling Maoris of long ago — yet they came and settled in this area.

iii. The First Comers

Driven either by the adventurous spirit of exploration inherited from their forebears, or by the urgent need to avoid annihilation by almost constant intertribal warfare, or by the apparently almost as strong desire to seek revenge; certainly from the necessity to seek more abundant food supplies or simply from a desire to see

what was "round the corner", the result was that there were few parts of New Zealand the Maori failed to penetrate — excepting only the great mountain ranges. Even these were traversed by wellknown routes over the major passes that, from one end of the Alps to the other, gave access to the West Coast and the main sources of supply of the indispensable greenstone.

The Wakatipu Basin was traversed by two main routes — that from the south, which followed the Nevis Valley down to where the now-collapsed Natural Bridge provided a hair-raising crossing of the Kawarau River; then, either by the Arrow and Motatapu, or the Cardrona Valleys to Wanaka, connecting with the Haast Pass route. The other route led up the side of the lake to the Routeburn Valley and the Harris Saddle, where it joined the route from the south via lake Te Anau to Martins Bay.

At one time branches of two tribes lived beside the lake. The Ngati Mamoe had five pas; Te Kirikiri occupied the site of the Park at Queenstown; Oteroto on the north side of the Kawarau Falls; Tititea on the south side; Takerahaka at Kingston, and one called Tahuna, probably located on the Peninsula, site of the present golf course. The Waitaha lived at the head, or Glenorchy, end of the lake, where greenstone of less useful quality was to be found.

Sometime in the 1700s, the isolation in which these people had sought to find peace had instead placed them in a perilous situation. Isolated as they were from the stronger units of their tribes, they became tempting targets for their current enemies, the more northern Ngai Tahu. One raid on the Tititea pa by a 300-strong taua was repulsed and the raiders were forced to fire the country to make good their retreat; but other raids were to follow and the threat was always present.

Whether it was due to this untenable position or that the tragic measles outbreak of 1835 struck even this isolated area, or whether the severity of the winters at last became unbearable, no Maoris were permanently residing in the area in 1853, when Lake Wakatipu was first seen by a European.

iv. The First Europeans

Nathaniel Chalmers, with two Maori guides, Reko and Kaikoura, was attempting to make his way from Southland to Canterbury by the inland route known to the Maoris, i.e. by Nevis Valley, Lakes Wanaka and Hawea and the Lindis Pass, when he had a distant sighting of the lake from a high point of the Slate Range near Garston. Chalmers was the first European to cross the Natural Bridge on the Kawarau — he reached Lake Hawea but, exhausted, called a retreat and the party rafted down the Clutha to safety.

The shores of the lake were first reached, from the south, by

Rees's and von Tunzelmann's first glimpse of the Wakatipu area would be almost identical with this view from the 3,677 foot Cardrona saddle. On the slope above the road may be seen "the insignificant yet appropriate memorial" — the pack track of the pioneers.

John Chubbin's party in 1856 — by this date practically all of the easier pastoral plains country of the South Island had been taken up under licence and it was to these remaining mountain lands that determined men turned, seeking fulfilment of their dreams of grand hills for sheep.

By 1859 at least seven parties had visited the south end of the lake, but were deterred from further advance by the precipitous mountains that in places plunged almost straight down into its waters. Of these parties D. A. Cameron and A. A. Macdonald probably penetrated the farthest, and secured licences for the gentler southern portion.

If he was not the first to enter the district, it was in this same year that one Donald Hay showed great determination and courage by traversing the lake on a korari (flax-stick) raft and exploring on foot much of the surrounding area. On applying for a licence he was told this area had already been spoken for and he returned in disgust to Australia.

At this period land speculation was common, many of the applicants for pasturage licences never having seen the land applied for. The land authorities, busy with the orderly settlement of the easier areas, simply did not have the time or the resources to adequately oversee the issue of licences, and often granted them in accord with the information supplied by the applicant, with generally indefinite boundaries. However, the "men were sorted from the boys" by the requirement that stocking of the land applied for had to take place within a stated time, failure involving cancellation of the application and the loss of an often substantial deposit.

Early in 1860 W. G. Rees and N. von Tunzelmann entered the Wakatipu Basin via the Cardrona Valley — the first successful attempt from the east. They made their first night's camp where Arrowtown now stands, their second at the Kawarau Falls (after a harrowing journey through the dense matagouri and speargrass on the flats) and their third in Queenstown Bay. From here, by raft and on foot, their horses being left hobbled at the bay, they struggled on as far as a high point past Bobs Cove, from which they could view the upper arm of the lake. On the return journey Rees set fire to the scrub with a view to providing easier access for stock at a later date. This fire secured such a hold that they were barely able to race it back to the bay, collect the horses and press on till the Shotover River lay between them and the fast-advancing wall of flame.

Rees was an energetic and capable man. Three weeks after his return to Dunedin on 14 March he had lodged and had had accepted the necessary applications; by mid-October he had transported a whaleboat from Invercargill to the lake and carried out a thorough exploration; by December the 3000 sheep he had had the foresight to acquire were on the move to the Wakatipu. The end of January 1861 saw the Queenstown run (no. 356) stocked and permanent quarters nearing completion in the bay, pending the arrival of Mrs Rees.

In 1861, with the Glenorchy run (no. 346) also stocked, the affairs of Rees had reached some stability.* Then gold was discovered.

Thousands of miners flocked to the area; a township sprang up, almost overnight, round his homestead in the bay, even his woolshed was pressed into service as an hotel. Rees's run was declared a goldfield and he was awarded £10,000 compensation, but was not permitted to retain any land, even in the bay where his homestead and buildings had been established. He was forced to move to the two other runs, nos. 345 and 331, which he and his partners, Grant and Gammie, had acquired.

These runs lay along the eastern side of the south arm of the lake and down the Kawarau River — they became known as Kawarau Falls Station.

* This run apparently remained stocked for only six months.

For Rees, the loss of the Queenstown run must have been a big blow and now, with the termination of his agreement with his partners drawing near, he had to face once more the problems of establishing a new station. Already the rumblings of a boundary dispute were evident.

v. The Early Surveyors

As mentioned earlier, circumstances had necessitated the issue of early pastoral licences on information supplied by the applicants. This had naturally resulted in misunderstandings and in some cases litigation, for the true location, let alone the actual boundaries, of licences granted, had no factual basis.

Each province of the time had its own surveyors and each worked on its own system — the survey work of the emerging nation was a shambles and that of the Otago Province was no exception.

This country is indebted to the action of the then Superintendent of Otago, Captain William Cargill, for appointing, in 1856, J. T. Thomson, a surveyor and engineer, to the position of Chief Surveyor of Otago.

Thomson was a man of exceptional ability. He had perfected a system of survey using solar observations upon which baselines could be stablished and from which further accurate survey could be extended. Later, as Surveyor-General, he was able to establish the system as standard procedure for New Zealand.

To further the work he had already carried out, Thomson engaged other surveyors, training them in his system. One of these was James McKerrow from Scotland who proved equal to the task of carrying out the survey of the some 3000 square miles of the mountainous country of the western part of the province.

Based in Dunedin, he set forth on several long expeditions into the interior, usually with two companions, three hacks and a pack-horse. The two all-important chronometers were carried by hand, and as can be imagined, proved to be "most wearisome burdens".

The men were able to obtain some supplies from the runs they passed through but lived to a large extent off the land, the abundant weka providing a ready source of food as well as being a great nuisance, from its penchant for thieving anything movable from their camps.

McKerrow's reconnaissance survey of the Wakatipu area was carried out during 1862-63. The first survey of the township of Queenstown was carried out by W. C. Wright in 1864.

A triangulation survey was carried out in the Wakatipu area by F. Howden in 1866, but it was not until 1909 that the actual boundaries of Kawarau Falls Station were defined and established by S. T. Burton.

Duckshooters at Kawarau Falls, 1902. Reverse photograph for reflection of Remarkables.

vi. The Owners of Kawarau Falls Station

The following is a list of the occupiers of the station — originally named by W. G. Rees as the Falls Station and known by that name to those associated with it; it later became officially known as Kawarau Falls Station.

Behind this abbreviated history lie the stories of the individual successes and failures of the occupiers, their personal problems and some tragedies, and their day-to-day struggle with an often uncompromising environment, in economic as well as natural terms.

The eastern mountainlands of the Wakatipu that were to become Kawarau Falls Station, in 1861 consisted of No. 331 (Staircase) and 345 (Peninsula). Run 331 was first licensed to Donald Angus Cameron in August 1860, but because of heavy losses in a shipment of sheep from Australia, Cameron was unable to fulfil the stocking requirements of all his applications and concentrated his sheep on the Nokomai licence. The licence for Run 331 was transferred to John McIntosh and Angus MacDonald, who must have sold within a few months to Grant, Gammie and Rees.

The considerations for which these and other runs were transferred at this period varied considerably according to circumstance. Cainard run was once transferred for the sum of one shilling; Mount Dewar run, known as Saddle Hill, was transferred for the sum of one pound with settlement in two payments of ten shillings each. Cameron transferred 331 in consideration of 300 ewes delivered to Nokomai.

Run 345 was first applied for by William Saunders, who was lessee of Waipori Station. It would appear that Saunders, following the practice of the time, had applied for country which he had never seen, but was described as being to the west of Shennan Bros. application (Big Kawarau). This licence, 345, was secured by Rees on behalf of the partnership and so the Falls Station was created.

The area of the licences held by Rees and his partners* in 1862 can only be assumed, for the boundaries had not yet been established — with 331 at 50,000 acres; 345 at 35,000; 356 at 100,000; and 346 at 54,000 the area would not be less than 239,000 acres and probably much more. Run 356 was the hub of this great and promising enterprise and when it was confiscated and declared a goldfield the structure collapsed — Rees was left to rally the partnership fortunes on the Falls.

It would appear that the partnership of Colonel William Lewis Grant, George Gammie and William Gilbert Rees was formed during 1858 (when Rees returned to England from Australia) for a set period of time, for the partnership was officially dissolved "by the effluxion of time", 27 July 1865.

Colonel Grant. This man remains a largely shadowy figure from the past, despite exhaustive enquiries by Mr P. Chandler, historian. It would appear he was a man of considerable economic substance, moving in the same social circles in England as did George Gammie. The author believes he must also have had interests in Australia about this time, for he is recorded as having visited this country on partnership business on four occasions between 1859 and 1871.

George Gammie. Son of an Aberdonian doctor. With his brother John landed in Australia about 1838. They arrived with two shillings and sixpence, won by John in a shipboard wager. In 1840-41 we find them trekking north with 3000 "catarrh-stricken sheep received in lieu of wages" to settle, or rather squat, on an abandoned run. After years of hardship they eventually prospered, and when John Gammie died in 1854 they were shearing in excess of 60,000 sheep. George sold the Australian properties, returned to England and purchased Shotover House, near Oxford. Here he established himself with a retinue of ten servants and

*Alfred Duncan, who accompanied Rees on the first drive of stock to the district, was a partner in the stock.

engaged in extensive property speculation at home and abroad.

His fortunes took a downward turn from about 1860 and would in no way have been assisted by the effect on his New Zealand interests of the rabbit plague and scabies disease in sheep, not to mention the disastrous snowstorms of the period. In 1865 he changed his name to George Gammie Maitland and in 1867 became bankrupt.

The decline in the fortunes of this gentleman adventurer, whose courage we can only admire and who, at the time of Rees's return to England was described as "a landed proprietor and magistrate", were to have repercussions on Rees's fortunes in far-off New Zealand.

William Gilbert Rees. Born 1827 into a naval tradition (his father was a lieutenant, who died at sea when Rees was twelve years old), attended a naval college; was an apprentice engineer then took up schoolteaching. He emigrated to Australia in 1852. In 1853 he became manager of Stonehenge Station on the Darling Downs where his neighbours were the Gammie brothers. This may be accepted as the beginning of a connection which finally resulted in his becoming the working partner on the Grant, Gammie and Rees holdings.

He returned to England in 1858 and married his nineteen-year-old cousin Rebecca Frances Gilbert. Rees and his bride sailed for New Zealand on the ship *Equator* which reached Wellington, 22 January 1859. He immediately set about acquiring land (on behalf of the partnership), in the South Island, where occupation was far from complete. Ahead of him, as we have already seen, lay three exciting and arduous years.

With the loss of Run 356 Rees moved to the Falls. The contract with the partnership which expired in 1865 was not renewed and the title passed to Grant and Gammie (now Maitland) who retained the property for only a year.

The Falls was sold by auction in August 1866 and purchased by Boyes Brothers, the sale agreement listing Runs 331 and 345; 7,138 sheep; 7 horses; 39 cattle and 3 working bullocks for a total price of £15,500.

Rees must have been retained on the station in an advisory, if not managerial capacity, as a stock mortgage of 1866 states that "the sheep were under the proper care and management of William Gilbert Rees".

In 1867 Rees left the Wakatipu district to become manager of Galloway Station for Campbell and Low, then a stock inspector in various districts, finally to die in Wairau Hospital, Blenheim, in 1898.

It would appear that his partners were unable to pay the money that was due to him at the dissolution of the partnership.

The *Boyes brothers.* Frank Campbell, Charles Crofton and George Henry (Captain RN), were the sons of John Boyes of Hobart, Tasmania, and his wife Sabina, daughter of George Meredith, Cambria, Great Swanport, Tasmania. (George Meredith claimed descent from Owen Glendower, "Prince of South Wales in the reign of Henry IV".)

Of his descendants, the Boyes brothers, George Meredith Bell and George Farbrace Boyes Poynter, cousins of the Boyes', were all associated with southern runs. George Meredith Bell was involved in the ownership of the Falls in 1879-81, when the Boyes' fortunes were at low ebb. George Poynter, from 1865 to 1870 a partner in the firm of Poynter & Butler who owned the Plains Station, Mossburn, also bought in 1870 from the Bank of Australasia (as mortgagees of George Gammie Maitland) the Dalvey run at Tapanui which had been secured for the Grant, Gammie, Rees partnership by Rees in 1859.

Charles Crofton Boyes was the resident working partner. Frank Boyes spent much time abroad, while G. T. H. Boyes was a career naval officer.

The Boyes Bros. ownership saw a rapid increase in flock numbers, reaching a peak of 29,000 sheep in 1877 — however, the 1878 snow and increasing problems with rabbits gradually reduced the flock to approximately 6000 sheep. F. C. Boyes withdrew from the partnership in 1875. In August 1886 the station was abandoned to the mortgagee, the New Zealand Loan & Mercantile Agency Company, which retained possession until 1898.

New Zealand Loan & Mercantile Company. They held the station with a succession of managers, namely: Donald Manson, William Menzies, and William Sams, each of whom struggled on, under the desperate economic circumstances of the times but in 1894 not only the Company but the Bank of New Zealand as well, was forced into bankruptcy. The Company (and the Bank) were refloated under the same names with new directors. The Falls lost all the Gibbston country to Big Kawarau (held by the New Zealand & Australian Land Company) and in 1898 was sold to Dan McBride.

Dan McBride. The McBride family came from County Antrim, Ireland, in the wake of the goldrush to Queenstown in 1862-63. Dan McBride and his brothers established what was to be a successful sawmilling business near Kinloch, Glenorchy. The mill had what has been claimed to have been the largest waterwheel in New Zealand. Unfortunately two serious bush fires occurred; the first, in 1887, was lit by stock thieves, and the second, in 1894, when a patch of bracken was set alight to dislodge a recalcitrant bull — the sawmilling, as a major industry, was ended. Dan McBride died in 1913 and the property passed to his son.

"With a steady snorting from the horse the crossing was soon made." The main point of access to the station.

J. P. McBride. John McBride guided the station through the difficult period of the First World War, in 1922 selling the property to Dickson Jardine of Glencoe Station, Crown Terrace, Arrowtown.

Summary.

Grant, Gammie & Rees	1860-1865
Grant & Gammie	1865-1866
Boyes Bros.	1866-1886
New Zealand Loan Company	1886-1898
Dan McBride	1898-1913
J. P. McBride	1913-1922
D. Jardine (family)	1922-present day (1977)

vii. Dickson Jardine & Mary G. Grieve

Dickson Jardine owned Glencoe Station, near Arrowtown, and ran The Falls in conjunction with that property.

When not yet out of his thirteenth year he rode up the Waitaki, his worldly possessions in a swag in front and a faithful old handy dog trotting along behind. He was headed for the big stations of R. Campbell & Sons in the upper Waitaki, to spend many of his early years on Benmore Station, which carried over 100,000 sheep.

As he met the squealing bullock waggons hauling out the wool, the boy would have had little other than his determination and his dreams to suggest that he was to become the manager of R. Campbell & Sons, big Otekaike Station, before it and Benmore were subdivided and sold, or that he was to become one of the recognised leaders in the industry, representing it on two Royal Commissions.

One warm Sunday morning at Otekaike, Dickson watched with some concern as a pretty young lass, Mary Grieve by name, a visitor from Otahuti in the far south, was being helped into a gig with two in hand to go for a morning drive.

The horses were by no means quiet. Fit as fiddles they reared and trampled about, requiring a man at each of their heads to restrain them. Dickson was considering making the suggestion that he should accompany the girl, as she settled in and gathered up the reins, but Mr John Tait, her uncle and the then manager of Otekaike, appeared quite unconcerned — and he, also, need not have worried.

When the horses' heads were released they took off with a scrambling leap, whirling out round the drive in a shower of gravel. As they were expertly straightened up for the gate, out came the whip, to crack crisply on either side of the leader's ears — for half a mile they went at full stretch, then she pulled them firmly in to a more sedate trot. In the cheer of appreciation let out by the men, Dickson's was probably the loudest.

He and Mary Grieve were married in 1909.

George and Rachel Jardine, with children Dickson and Ellen.

"A pretty young lass, Mary Grieve by name."

After their marriage at Meadow Park, Otahuti, Southland, in 1909, Dickson and Mary Jardine settled on the homestead block of Otekaike Station which had been subdivided in 1908. The station comprised 17,495 acres of freehold and 48,965 acres of pastoral country was subdivided into some thirty-four runs and sections.

Their first son, Grieve Templeton, was born at Otekaike in 1911.

The main homestead at Otekaike had been purchased for use as a Special School and, when Dickson was away managing Benmore Station for a short time, one of the inmates made a night attack on Mary, alone with her young son. Although she was not physically injured in any way, this was to result in an illness that was to last nearly a year; it turned her hair snow-white almost overnight.

Dickson and Mary Jardine in 1919 with Grieve, their eldest son and the author.

They sold Otekaike shortly afterwards, in 1913, moving to Invercargill to encourage Mary's recovery. Despite this setback, between 1913 and 1922, when they finally settled on Kawarau Falls Station, they had purchased, improved and resold five stations, namely Diamond Peak, Fairlight, Birchwood, Sharpridge and Glencoe.

Mother ("Chick" to her friends) was a wonderful person. From their first property, the Otekaike homestead block, through all the ups and downs, the cooking for gangs of men and the seemingly continuous shifts from one station to another, she remained ever cheerful.

An expert organiser, she was also an extremely capable artist in oils, watercolour, pokerwork and — butter. She not only made the butter but, if there were visitors coming, would create a centrepiece of roses and flowers for the table out of butter that were perfect reproductions, all in a matter of minutes. The support she gave her man is typical of the very considerable part played by the womenfolk of the high country.

During this period Dickson Jardine served on the Royal Commission on Southern Pastoral Runs — he was to serve on a further Royal Commission in 1939.

The second son (the author of this book), Dickson Glendinning, was born at Fairlight Station in 1916.

With Glencoe sold in 1928, Dickson Jardine concentrated on Kawarau Falls — a partnership with the two sons was formed in 1941 and in 1947 the property was divided between them, Grieve taking over the homestead block and Dickson G. the high-country section. Grieve sold his portion of the station in 1960, moving to a farm in Southland. He died in 1973.

Dickson Jardine died in 1952 and Mary in 1960.

viii. Pioneer Forbears

Dickson Jardine:

His parents, George and Rachel Jardine of Dumfriesshire, emigrated from Scotland, sailing on the maiden voyage of the *William Davie* on 1 August 1867. They joined George's two brothers, already settled on Annandale farm near the Waitaki River, north of Oamaru, where Dickson was born in 1878, the last of a family of eight.

My mother's parents emigrated from Scotland on separate occasions. Her father, John Glendinning Grieve, came out with his parents, John and Mary Glendinning Grieve and other members of the family, sailing on the *Strathmore* from London to Dunedin in 1856.

Robert Grieve, the eldest son, had already been in New Zealand for six years and had selected land for his father near Invercargill, naming it Branxholm, after the family farm Branxholm Braes, Hawick, Roxburghshire. This was the first freehold land in Southland, being Section 1, Block 1, New River 100, and when Invercargill was surveyed two and a half years later the survey had to begin from Branxholm. Robert shore and dispatched the first wool from Southland.

After arrival at Dunedin four of the brothers set off in February 1857 with a packhorse and travelled overland to Branxholm to

John Glendinning Grieve and Helen Grieve (née Tait), 1872.

prepare for the coming of the family, the journey taking five days.

They engaged two sawyers to cut timber and erected a comfortable hut for their parents and other members of the family, who arrived in April, having sailed in the famous schooner *Star,* the only coaster plying between Dunedin and Invercargill, to the New River harbour, then by open boat up the Oreti and Makarewa Rivers to the Tomopurahau Creek.

"John G. Grieve left home to try his luck at goldmining when the Gabriel Gully rush broke out in 1861. He was unsuccessful there, and after a time, when the West Coast rush took place, he made his way there. Leaving Dunedin in a small boat the party eventually reached Hokitika after an extremely perilous passage, for the craft was outrageously overloaded. On arrival the pilot demanded £300 to take their boat into the river. The captain, thinking the amount was out of all reason, determined to attempt it himself, with the result that he missed the river and landed the boat in the surf. After an exciting time much of the provisions and a number of the swags were lost and the result was much hardship. John was decidedly unsuccessful here as well and, just as he seemed to have struck the spot where the golden nuggets lay, he was crushed by a fall of earth and consequently had to lie up in a tent for a considerable time. When he recovered he decided to quit goldmining and left for home on foot, making his way across country to Christchurch thence down the coast to Branxholm, arriving back about 1865. He then engaged in waggoning wool with a team of bullocks from Sunnyside Station up the Waiau to Invercargill. After this he returned home and took charge of the cattle, leasing land for grazing purposes in different parts of Southland.

"He became manager of the New Zealand & Scottish Investment Societies' Spar Bush cattle run of 5150 acres until it was cut up. In those days of wide open spaces with large mobs of cattle and no fences, John G. Grieve earned a reputation for himself as a stockman. The wild outlaw bullock that could not be brought down and hogtied by 'Waimatuku Jack', just did not exist." (*quote S.T.*)

In 1870 John G. Grieve took up a Spar Bush block and after building a house and outbuildings married Miss Helen Tait of Waianiwa, at Jerviswoode in 1871. Four years later he purchased 621 acres of virgin land at Otahuti, naming it Meadow Park.

Shortly after the Grieve family settled at Branxholm a tragedy occurred that was to cast a shadow over their early years in their new homeland. In these early days, the only minister available to perform the rites of marriage was the Rev. J. F. H. Wohlers who resided on Ruapuke Island — lying twelve miles off the Bluff and some forty miles from the New River harbour. Any voyage to the island across the turbulent waters of Foveaux Strait was not to be taken lightly.

"On 8 December 1858 Adam Wilson and Margaret Grieve, John G.'s sister, set forth to be married. They left Invercargill, in the New River, in a ten-ton boat owned and skippered by George Hunt and and crewed by Robert Smith and accompanied by John Jackoways and his wife, the former Kitty Kelly, a daughter of Invercargill's first townsman.

"The weather was worse than anticipated at the heads and when crossing the bar the boat struck a sandbank and filled — no boat could have reached them and they were all drowned, only the bride-to-be's shawl coming ashore. When the seas calmed some Maoris took a boat to the scene of the tragedy and were terrified when they saw Kitty Jackoways standing upright on the bottom, her dark tresses streaming in the tide, her feet having become embedded in the sand." (*quote, Historical Southland*)

Mary Grieve's mother, Helen Tait, at age fourteen years, emigrated with her parents Robert and Margaret Tait and other members of the family, from Glasgow to the Bluff in 1863 — they came from Sneep, the family farm at Kelso, Berwickshire. The voyage took 156 days, during which they made only one port of call, at Cape Town, when they ran short of water.

Robert Tait was appointed manager of Clydesdale Station on his arrival in New Zealand, but having a young family requiring education, settled in Waianiwa, Southland.

My wife is descended from James Reid, eldest son of a family of four, of Hunton farm, Stronsay, Orkney Islands. In 1857 it was decided to investigate conditions in the colonies and, as was the usual practice at that time, the two younger brothers were deputised to go out and report back to the family. They spent some time in Australia, then came to New Zealand in 1861, landing at Dunedin. Their reports of New Zealand encouraged the remainder of the family and some relatives to emigrate to this country.

James and Jane Reid, their two infant sons James and Peter, and other relatives came to New Zealand in the barque, *Dona Anita,* landing at the Bluff in January 1864.

James Reid mined for a short time with his brother David at Arrowtown (his brother Peter having been killed in a mining accident at the Sowburn, Central Otago) and he then took up farming and contracting in the district. Following the practice of the Old Country, he built his house in the village of Arrowtown, working his farm, which adjoined the village, from there. James and Jane eventually had a family of eight.

James and Jane Reid, *circa* 1885.

My wife's father, William Stewart Reid, the second youngest son, joined his elder brother Peter on his farm Glenpanel, Lower Shotover, after leaving Southland Boys' High School. Peter had extensive mining interests in both Australia and New Zealand and the younger brother conducted the business of the farm. William then purchased four deserted farms, combining them to form Willowbrook, Arrowtown, and married Mary Georgina Martin in 1914.

Her father, John Martin, was born at Saintfield, Northern Ireland in January 1834 — he had one older brother James, grandfather of Sir James Martin, inventor and builder of the famous Martin-Baker ejector seat that has already saved the lives of over 4130 airmen.

The brothers lost their parents whilst John was still very young; he was taken to live with his grandfather at the home farm of Killinchy Woods, Crossgar, County Down, Northern Ireland.

John suffered from indifferent health as a youth; at 19 years of age, on the advice of his doctor uncle, he took ship to Australia; benefiting from the climate, he and a partner took up a contract to cart poles for the great Overland Telegraph. During this time he had the doubtful pleasure of meeting Ned Kelly the bushranger.

With the completion of the pole contract he turned to gold-mining, coming to New Zealand in 1860. He reached the Wakatipu district among the first of the waves of goldminers and, with a partner, George Harrow, took up and worked an alluvial claim near the outlet of Bracken's Gully on the Arrow river. He also had shares in the famous quartz mine, the Homeward Bound. During one very bad winter he undertook to distribute milk to the children of miners at Macetown, nine miles up the river from Arrowtown.

He returned several times to Northern Ireland during these early years; in August 1877 he married Sarah McCammon Gourley of Ballymacreely farm, Derryboy, County Down. They returned to New Zealand in the sailing ship *County of Peebles,* arriving at Port Chalmers in February 1878. They came on to Arrowtown where they lived for a time, whilst John awaited the surveying of his and other farms on the Crown Terrace. With the completion of the surveys Sarah climbed the recently finished Tobin's Track from Arrowtown to their Lilburn farm on the Terrace, carrying her first-born son James on her back. Here her home was a tent, until such time as their permanent home was built on the edge of the Terrace, with a commanding view of almost the whole of the Wakatipu basin.

John was a good farmer and a public-spirited man — a sample of the wheat he grew on his Crown Terrace farm took first prize at an exhibition in Melbourne, Australia. This may seem strange

John Martin, 1877.

Sarah McCammon Martin, 1877.

to the modern reader, but at that time a regular, and popular service ran between the Bluff and Melbourne. He was one of the five Terrace settlers who, in 1897, contributed funds for the establishment of a school on the Crown Terrace; he was also chairman of the first school committee.

The Crown Terrace school opened with a roll of 21 pupils — prior to this the children had to walk to Arrowtown, a distance of three miles involving a descent of 1000 feet, and, of course, a similar climb on the return journey. When the Terrace school was opened it attracted children from the Arrow Junction, who made the walk, or rather climb, from there to attend.

The school remained open until 1957.

John Martin died in 1905 and Sarah Martin in 1947. They had a family of six — Mary Georgina, the second youngest, marrying William Stewart Reid on 28 October 1914.

They were married in Arrowtown and the reception was held at the Martin family farm on the Terrace. Unfortunately a severe storm struck the area over night bringing a fairly heavy snowfall, which played havoc with the big marquee that had been erected for the occasion. The bride was transported in the car belonging to the best man, Peter Reid, over the Zigzag road, Tobin's Track not yet being opened to motor traffic. He also transported the bride and groom to the Antrim Arms Hotel at Frankton, no motor traffic being allowed past that point, from where they caught the horse coach connecting with the lake steamer at Queenstown the following morning.

In addition to William and Mary Reid's farming operations at Willowbrook, in 1919 the 100,000-acre Branches Station at the head of the Shotover River was purchased, where Hereford cattle and sheep were carried — the wool being packed out the twenty-four miles down the river and over the 3000-foot Skippers Saddle. This property was sold in 1955.

William and Mary had a family of three, Jessie Stewart, John Gourley and Alan Lindsay — they all took an active part in the running of the properties.

Jessie Stewart Reid and the author were married in 1940. They have two sons and one daughter. Dickson Stewart and Andrew Glendinning are both married and in partnership with their father on the station, and Janetta Mary is now Mrs Wynn, also living in the district. There are, to date, five grandchildren.

ix. Kawarau Falls Station

On 11 April 1923 Mother packed their personal effects for what was to be the last shift — from Glencoe to Kawarau Falls.

In that daunting undertaking of shifting house, she had by now

Sculptured by Mary Jardine using butter, a piece of cheesecloth and a pair of butterpats — "Roses and flowers that were perfect reproductions". First prize entry — the scene at left is on a mirror.

become expert. Goods and furniture were soon loaded on Dave Mackie's waggon for the trip down the 800-foot drop and hairpin bends of the Zig Zag, over the easier going flats and the long stretch of the Ladies' Mile to the Frankton Arm of Lake Wakatipu.

It was a featherweight load for the powerful, fast-stepping team, but on this day, what with one thing and another, it was dark when they arrived and the job of transporting the furniture across the lake in the rowboats had to await the light of the following day.

Here, where the Frankton Arm narrowed sharply, the apparently still waters of the lake sprang abruptly to life, sliding swiftly down to go rumbling and leaping through the rocks of the Kawarau Falls.

The homestead was sited a little over 200 yards upstream from the falls and 100 yards back from the lake. It was backed by a large plantation of pine and gum trees, surrounded by extensive lawns and gardens, with the lake-front edged by a noble line of huge but graceful weeping willows.

Commenced by W. G. Rees in 1863, with additions by successive owners, set in approximately seven acres of gardens, orchards and yards; backed by a large plantation of trees, this was a gracious home.

The wooden, partially two-storeyed house had the long passaged, rambling, but none the less homely and characterful interior of the houses of the era. Two large verandahs were included on the north and west sides. Hewn from the local rock the three large front steps made a solid entrance.

Seven open fireplaces and a coal range provided heating and cooking facilities and lighting was by kerosene lamps and a rather fickle electric power plant.

Being then a growing boy the author has vivid memories of the kitchen. With bins and cupboards ranged round the walls, the coal range at one end and a large table and forms on one side, it had a ceiling centrepiece of a long, pull-up clothes dryer, on each side of which hung rows of home-cured bacon — this was one of the busiest places on the station.

The rather antiquated plumbing system caused many a headache — subject to an almost continuous, if spasmodic, process of updating, it was prone to freeze up in the most awkward places. The muffled roar of a blowtorch was no uncommon sound during winter months — especially was this so during the winter of 1923.

Only a few months after moving into her new home, the author's mother had to contend with what was to prove to be one of the worst winters recorded in the district — over four feet of snow fell in some parts and thousands of sheep were lost. Some owners of stations in the area never recovered from the blows dealt by the storms.

Two feet of snow fell round the homestead and lay for weeks; half-hearted thaws were followed by raging blizzards that laid down more snow. Temperatures dropped to record lows — so low the large plantation of gumtrees behind the house were all killed. Trees sixty feet high and over two feet thick were frozen solid and burst with the frost — these explosions during the long still nights sounded like intermittent gunfire. Most of the poultry died, frozen solid to their overnight perches.

These gumtrees have, as has been witnessed down the years, great powers of revival — from seed and living butts they spring again to life, to eventually hide the still standing skeletons of their fellows.

On the two stations, the menfolk were fully occupied in fighting to save what stock they could, and there was little time to spare for such things as frozen pipes at the homestead.

The rather complicated roof that had evolved through the various additions, held the snow like a bucket — no snow-boards had been provided and the gutters quickly blocked with frozen snow; each thaw sent water cascading down into passage and room, to freeze on the floor where it fell. The long passages became skating rinks, despite vigorous mopping up, and improvised braziers were pressed into service in an effort to thaw and dry the house.

Despite the running battle against leaks, frozen pipes and a small boy who was one big chilblain, there was always a hot meal for the dog-tired and dispirited men who would appear unexpectedly from days of snow-raking.

Fortunately, seasons change, and the large garden and grounds begun by Boyes Brothers in 1866, posed a challenge the author's mother could not resist — the fruits of her planning and labours were to stand for many years. Rustic fences, shrubberies, rockeries, rose gardens, etc., not to mention a vegetable garden that would not have disgraced a market garden — all these came into being as the years passed. She not only created a beautiful steading, but above all, a home — unfortunately, only a house now remains.

This was the main point of access to the station — by rowboat. The only land access was by pack track, up the side of the lake from Kingston or along the south bank of the Kawarau from Gibbston.

All farm produce was shipped out on the lake steamers from the Woolshed Landing, six miles to the south. Halfway between the homestead and the woolshed, in the valley at the foot of the Remarkable Range, was sited the working centre of the station — men's quarters, cookshop, stables, buildings and yards, known as the Dip.

When Rees and his partners owned Kawarau Falls it had an area in excess of 100,000 acres and in latter years, before the rabbit

Kawarau Falls Homestead — commenced 1863. The photograph was taken in 1929.

plague, carried a maximum of 29,000 sheep. At the time my father took over the property it had an area of 42,000 acres, the maximum carried in his time being 11,000 before subdivision within the family.

Glencoe and The Falls, combined, carried a maximum of 19,000 sheep and 300 head of cattle. Glencoe was sold in 1928 and my father directed all his energy to the development of The Falls.

Bad snows occurred in 1923, '33, '39, '44, '47 and '68.

The introduction of rabbit boards in 1949 saw the beginning of control of the rabbit, exactly seventy years after the first attempt at organised control. The success of the boards, coupled with a comprehensive aerial oversowing and topdressing programme in the 1960s, created conditions suitable for the introduction of cattle.

The two events having the greatest long-term impact on the station, both for good and ill, were — first, the opening of the Kawarau Dam in 1926, which, as long ago as 1889, had been advocated by Sir Julius Vogel. As well as being an efficient dam,

it is probably one of the strongest traffic bridges in the country. Then, in 1936, the lake road to the south was finally completed. In the intervening years it has grown from a mere access track to a high-speed highway that has slashed open the natural boundaries of all the various blocks and placed a heavy burden on the property.

Today, the station, very much reduced in size, still carries some 13,000 ewe equivalents — the name has been changed, but its heart, the Remarkable Range, remains.

x. The Remarkable Mountains

When Alexander Garvie was carrying out a reconnaissance survey in the Maniototo area of Otago in 1857-58 he saw, far to the west, a compact group of high snow-covered peaks, and named them the Remarkable Mountains. James McKerrow in his survey of 1862-63 overlooked Garvie's original name and called the whole range the Hector Range. This duplication was overcome by adopting the name Hector Mountains for the larger area and retaining the name of Remarkable(s) for that portion bounded by the left branch of Wye Creek and the Rastusburn, facing west towards the lake.

In 1864 T. N. Hackett and party were the first to climb to the top of the Remarkables.

Garvie could not have known how apt the name he had chosen was to prove to be, for the western face is truly remarkable. Soaring upward with dramatic abruptness from valley, lake and plain, the jagged peaks, bluffs and spurs culminate at 7680 feet in what appear to be two peaks known as the Double Cone, but which are really a group of three. These were first climbed by Mr Hackett, Government Geologist, in early October 1864.

There are many larger and higher ranges, but few more unusual or strikingly beautiful. The morning sun of summer sends searching rays into the blue mists that veil the plunging bluffs, spotlighting a jagged peak here or a rugged bluff crowned with golden tussock there, until the full warm sun of afternoon reveals every spur.

In winter, the bluffs, too steep to hold the snow, range like black guardsmen up the snowfilled chutes. The peaks stand sharp

Hackett and some of the party can just be seen as tiny dots on the summit.

against a brittle sky and the slightest wind sends snow banners flying somewhere along the face.

Spring spawns the avalanches that thunder down the chutes until only scattered drifts remain, to bejewel the black of bluff and gold of tussock beneath the summer sun.

Moon-backed at night they stand, a wild black silhouette of primeval mystery, untold ages apart from the hurrying lights and traffic on the roads below.

It is not difficult to assume that it was round the Remarkables that the Maori of long ago wove the legend of the monster Kopuwai, the water-swallower (see Appendix I) who drank the Kawarau River dry — for, on a still day, though one can hear and see the water tumbling down the higher reaches of the creeks, none reaches the valley floor except during thaws or heavy rain. Also, down the Kawarau River, where much of the country is basically unstable, there is ample evidence of many past massive geological slumps and landslips that would have dammed the river for short periods. Within the period of European settlement several such slips have occurred, one major slump completely damming the river for several hours.

No forests excepting isolated stands of mountain beech now clothe these slopes, yet along the face of the Remarkables, at an altitude of around 3000 feet, lie hundreds of gnarled old totara logs confined within a narrow band only a few hundred yards across.

Samples of these, sent to the DSIR, elicited the reply that they were some 600 years old and formed the totara band in the forest that reached into these areas in early times and was destroyed by fire. In the report of the destruction of these forests by fire can be seen confirmation of the basic facts handed down in the old Maori legends.

However, despite scientific evidence there are several factors relating to this line of timber that have not yet been fully explained.

1. Why does it lie within such a narrow altitude band? 2. Why does it lie almost exclusively on southern slopes? 3. Why are there no stumps (at least on this country)?

One is tempted to think that Maui momentarily let New Zealand slip off the hook and that the timber lies on an old water level.

Originally, it was thought that the highest level reached by the water within the basin did not exceed 200 feet above the present lake level, but now evidence is being found that it reached a great deal higher. In time those who lean towards the above "drift-wood" theory may be proved right.

Nature seeks to reclothe these slopes. Tane, Maori lord of the forests, has not given up hope of regaining this portion of his kingdom, for in the past fifty years most of the beech stands have expanded considerably, in proportion to their size. This expansion is interesting, for it does not appear to be a continuous process but, in the larger stands, proceeds in about half-chain leaps every five to seven years.

In the right-hand branch of Wye Creek, at an altitude of 4000 feet, among huge boulders below thousand-foot-high bluffs, there is a large collection of moa bones. These were discovered by D. Rowe and W. Miller in 1927 and when I first saw them over forty years ago they were in sound condition, but they have since almost completely crumbled away. This group of moas (one hestitates to refer to such giant birds as congregating in "flocks") may have perished in an extra heavy snowfall, for I cannot swear I saw the marks of human teeth on the bones. They are a tiny part of the faint, elusive traces of the prehistory of New Zealand. Unfortunately, the most likely sources of evidence have been swept away, mainly uninvestigated, in the stress of an emerging nation.

xi. More Shadows on the Hill

In that portion of the fabric of history woven into this district from 1860 to the present day, we see the weft reflected in the bright colour of gold and the exciting hues of commercialism, interspersed with the more sombre hues of wars and depressions. The continuing factor, the warp, formed by the stations and farms, runs through all, for they are the foundation of this era of the district's history.

The methods evolved by our pastoralists in wresting productive return from our mountain lands find comparison only in small areas of the British Isles, and this is easily accounted for.

When the colonisation of this country first began, it was from the hill country of England, Ireland, Wales and more especially Scotland that the shepherds came, seeking to build a brighter future in this new land, in the manner they knew best. Some of their traditions of herding persisted for a few years but soon, moulded by necessity and circumstance, there evolved the present system of free-range grazing of mountain country, unique to New Zealand.

The pastoral industry of our high country was built through the efforts of the shepherd and his dogs. These early shepherds, whose names have gone largely unrecorded, were the true pioneers of our mountain lands.

What draws and holds men to work in what is, in fact, one of the frontiers of human endeavour? Here the stock graze no gentle sward, but harsh tussock on ledge and spur, the vagaries of Nature are multiplied a hundredfold, and real danger can place its hand on a man's shoulder several times in one day. Here lies no certain economic security other than what can be guarded by experience and wrought by physical exertion against the extremes of Nature.

Ask the mountaineer why he climbs; what he does for pleasure is work for the musterer. Ask a high-country shepherd and he will probably mutter something about it being "a good run of mustering", or that he is "just filling in time before he gets a job on the flats". You will almost certainly *not* be told that he loves the freedom of the high places, the satisfaction of working in unity with others to a common end, or of the interest and pride he has in his dog team.

The word "romance" has many shades of meaning. We see it applied to the sea, to the cattle ranches of America, even to big business, but it can in fact be applied to any human endeavour demanding more than run-of-the-mill courage and endurance. It can very fittingly be applied to the high-country industry, not alone because of the demands made on the human participants, but also because of the environment in which its activities take place, the ranges and bastions of the great Southern Alps.

It is a facet of our national life of which we can all be proud.

The saga of the high country, the economic use of New Zealand's mountain lands, begun by pioneers such as Rees, continues today but, influenced by national and world trends, there have been and will be changes.

For well over a century now, the mountain lands which border the main alpine chain have been under the practical, if not political, control of the pastoralist. The use to which he has put these lands has, generally speaking, left them in good heart — for they were, in most cases, his life, even his love. They were the base upon which he built a way of life, itself a prized heritage and often the only one he had to pass on to his sons.

The mountain lands were not exempt from the environmental mistakes, even disasters, wrought by a developing economy; these were not all, as some would have, due to the ignorance, avarice or carelessness of the pastoralist — much wider influences, both economic and environmental were at work, over which the pastoralist had no control.

Compared to other parts of the world, nature, secure behind the effective isolation of a thousand leagues of sea, must have had few problems in evolving a balanced environment in this land with its uncomplicated avian fauna. This environment that had evolved over untold ages must have been beautifully balanced — until the rising tide of European expansion and settlement broke on these almost virgin shores. It can be seen now that the impact was severe, the full effects of which will unfold with the passing years.

The growing awareness that man is not a separate entity to nature, but an integral dependant part; that the laws of nature are deceptively simple, but none the less unalterable; that he can no longer alter an environment with impunity or regard the earth "and all the fruits thereof" as subordinate to his requirements, but must protect and cherish to survive — these have made us look more closely at the conservation of our natural resources, of which our high mountains are now recognised as being an important part.

Have we in this land allowed control of our expansion to slip away into the vicious circle that demands more and still more electric power for spiralling population and avaricious secondary industry alike — must we flood still more valleys, lakes and rivers to appease the demand. The mountain lands' contribution to the national economy is seen to be more valuable in water than in wool — some look to the day when this is their sole function. It could well be that these people are like a horse with blinkers, unaware of all the side issues; the stewardship of the high-country runholder should not be lightly tossed aside.

The memories of the men, and the part they played in shaping the past of our mountain lands, are now but "shadows on the hill". Of one thing we can be certain, just as cloud shadows have drifted across these mountains for untold years beyond our ken, so too will they do in the future — and men with a love of the mountains will be there to play their part.

Appendix I

TWO MAORI LEGENDS

LEGEND OF THE SLEEPING GIANTS — LAKE WAKATIPU

IN the heart of Otago Central there once dwelt a chief who possessed a very handsome daughter named Manata. Many a young man longed to have her for his wife but the old chief was very particular about the choice of a son-in-law. It came about that a young warrior named Matakauri, unfortunately of low degree, fell in love with Manata, who reciprocated his affection. The old chief was furious and kept his daughter under close observation, but a giant named Matau came and stole her away.

The heartbroken father then offered Manata in marriage to the man who could bring her safely home. Strong young warriors quailed at such a task, but the love that throbbed in the heart of Matakauri was so great that no obstacle could stand in his way. He went forth on a preliminary scouting and discovered that where the giant Matau dwelt, when an east wind blew, the monster invariably went into the deepest sleep. Selecting a day with that favourable wind, Matakauri essayed forth. He found Matau fast asleep, but he also found Manata fastened by strong cords.

Manata's heart throbbed with joy at the sight of her lover, and Matakauri set to work with his mere to sever the bonds but, try as he could, he failed.

Manata then sobbed bitterly, and her tears fell on the thongs. So great was the love embodied in her tears that the cords melted away. Matakauri picked her up in his arms and carried her safely home to the old chief who, as a reward, allowed the lovers to become man and wife.

But while Matau lived, no maiden was safe, so Matakauri sallied forth once more on a day when the favourable east wind blew. The brave warrior found the giant sound asleep as expected, and set fire to the large growth of bracken the giant used as a bed. The giant was smothered in the flames, the fat from his body augmenting the fire until the blaze was so fierce that it burned a deep hole in the land. The snow on the hills melted, streams poured forth, and filled the hollow; you can see it today — it is Lake Wakatipu.

The above version of the legend joins the sleeping giant with the Mataura River. Others continue "that the waters extinguished the fire before it reached the giant's heart, which still beats deep below the surface of the lake''. This is the rise and fall referred to. All versions make reference to the fire — could this be some faint memory of a conflagration that swept away what forest there was round the lake?

By such easily remembered and interesting legends were basic historical facts passed down by the Maori.

LEGEND OF KOPUWAI

In the vicinity of Kawarau, near the outlet of Lake Wakatipu, dwelt a veritable monster named Kopuwai. One of his captives was a Maori girl named Kaiamio. On going to sleep the monster fastened her to his leg, considering that a tug now and then would assure him that she was still secured. When the monster eventually went to sleep, she substituted a large log of wood in place of herself. Kopuwai tugged occasionally and, feeling the strain, considered all was well. The girl however had obtained a flax raft and escaped down the river to her friends at the sea coast. On waking in the morning the monster was extremely angry at being duped, and his wrath was so great that he swallowed so much water from the Kawarau that for a period the river emptied right down to its rocky bed.

APPENDIX II

PLACE NAMES

THE HISTORY of any area is in part recorded in its place names, but the origin of these names can be a controversial subject.

It is almost impossible in all except a few instances to state with any degree of accuracy who bestowed a particular name, whom it was intended to commemorate or, alternatively, the meaning it was intended to convey. These difficulties are accentuated by the passage of time and compounded by spelling alterations adopted through common usage — this is particularly so in the case of Maori place names.

The Maori had no written language and the efforts of the early European settlers, notably the church leaders, to rectify this, resulted in generalisations that of necessity could take little account of the varying district or tribal dialects. With many Maori place names this has resulted in uncertain or diverse meanings.

The following list of place names and their origins refer only to those areas associated with this book and are only as correct as are the sources of information from which they came.

Place	*Named after, or meaning*
Alta, Lake	Possibly named by T. R. Hackett, 1864. Meaning high.
Afton Peak Afton River	Named by Jas. McKerrow, 1863, after a place in the Scottish Ayrshire countryside.
Alfred, Mt	Named after Alfred Duncan, Rees's cadet. With G. Simpson he was the first shepherd at the Head of Lake. Named by his companions on the first expedition with sheep — G. Simpson, D. MacAusland and Bob Fortune (boatman).
Arrow, river Arrow, town	Possibly named after the river of the same name in Devonshire. Named by Rees in map supplied by him to J. T. Thomson, Chief Surveyor of Otago, 1856-76.
Arthurs Point	Named after Thomas Arthur, co-discoverer with Henry Redfern of gold in the Shotover River 1862. Named by common usage.
Ben Lomond	After the Scottish mountain. It was first bestowed on the eastern peak of the range, but the name was later transferred to the western peak. Named by Duncan MacAusland, shepherd for Rees in 1861.

Ben More	Scottish nomenclature, given by James McKerrow (?) surveyor, 1863.
Ben's Basin	After a huntaway dog belonging to A. S. Robins, which was lost in 1938. Named by common usage.
Bayonet Peak	After its appearance. Named by James McKerrow, surveyor, 1863.
Bowen Peak	After Sir George Bowen, Governor of New Zealand, possibly by the Govt. Survey Office, following the visit of the Governor to Queenstown, *circa* 1869.
Bob's Cove	After Rees's boatman, Bob Fortune. A favourite sheltering place. Named by common usage.
Brown's Basin	After early shepherd who kept a boundary there. Named by common usage.
Bucklerburn	After a farm in Aberdeenshire owned by Alfred Duncan's father. Named on first expedition with sheep to Head of Lake in 1861, by Alfred Duncan.
Cardrona, Valley	A Scottish place name, given by J. T. Thomson.
Cecil Peak	Named after Rees's eldest son, Cecil Walter, 1862, by James McKerrow, surveyor.
Coronet Peak	Named after a rock formation on the peak, by Rees, 1860.
Crown Range	Named after a group of rocks at summit by Rees, 1860.
Crichton, Mt	Probably after Geo. Crichton, T. Dalrymple's companion in the Wakatipu area, 1861. Named by James McKerrow (?) surveyor.
Dart River	After Dart River in Devon, England, by Rees, possibly on a boat voyage to Head of Lake in 1860.
Devil's Staircase	Named after the rugged nature of country.
Devil's Lumber Box	For the fact that lake currents deposit large quantities of driftwood there. It is uncertain who named it. Some authorities say Green, Rees's boatman, after taking a horse to Kingston in 1861. Others credit D. A. Cameron in his 1859 exploration.
Dip (The)	For association with a large sheepwash used in the early days of the station to clean fleece before shearing. Named by common usage.
Diamond Lake	Named for its beauty by Rees, 1860.
Elfin Bay	Earlier known as Sandfly Bay. Named by common usage.
Falls (The)	After the adjoining falls. Name given to station by Rees, 1862.
Frankton	After Rees's wife Frances (nickname "Frank"). Named by Sir J. C. Richardson, Superintendent of Otago, 1861. Surveyed, 1863.
Glenorchy	Scottish place name, in Argyll. Unknown. Surveyed by G. M. Barr, 1864.

Place	*Named after, or meaning*
Lake Hayes	After Donald Hay, who made the first recorded penetration of the district. There is no record of Hay having reached this lake, though he mentions another, probably Lake Johnson. Name given by Robert and Archibald Cameron during their exploration of 1859-60. They were nephews of W. Cameron, Athol.
Hector Mts.	After Sir James Hector, Otago Survey Dept. Named by James McKerrow, 1862.
Hope, Lake	After Scottish Loch Hope, Sutherland.
Hawkes Pass	After an early musterer. Named by common usage.
Kawarau River	Maori name of several meanings, including "bitter leaf". (A very old medicinal tree still stood in the early 1920s.)
Kingston	Probably named after Kingstown (now Dun Laoghaire), Ireland. Known briefly as St Johns after St John Brannigan, police commissioner, 1862-63.
Kirkpatrick, Lake	Named after James Kirkpatrick, first owner of Closeburn Station. Named by local usage.
Larkins, Mt.	Named after Sir John Larkins Cheese Richardson, by James McKerrow.
Lochy River	After a Scottish river, named by Donald Angus Cameron, 1859.
Lochnagar	Possibly named after a Scottish loch by James McKerrow or Charles Cameron, an early miner.
McKenzies Chute	After Duncan McKenzie, musterer, 1922. Named by common usage.
Macetown	After the Mace brothers, miners. Formerly known as the Twelve Mile. Named by H. Wilmot, surveyor, 1875.
Moke Lake	After a mule, "Donald", used by D. MacAusland in search of a route to Head of Lake. Named by D. MacAusland, 1861.
Morven Hill	Named after Morven Hill, Deeside, Scotland, by A. Duncan, 1861.
Moonlight	Probably after George Fairweather Moonlight, a miner, by local usage.
Mt Gilbert	Named, probably after the Gilbert family connection of Rees, by James McKerrow, surveyor, 1863.
Nevis, River Nevis, Ben	After Glen Nevis area and Water of Nevis, Scotland. Ben Nevis is the second highest peak in group of ranges. Named by Donald Angus Cameron, 1859.
Nicholas, Mt.	After Nicholas von Tunzelmann, companion-explorer of Rees and first owner of Fernhill Station, 1861. Named by James McKerrow, 1863.

Appendix II

Place	*Named after, or meaning*
Queenstown	Originally the site of "The Station", Rees's homestead, 1860-62. Known to miners as "The Camp". Named Queenstown (now Cobh), County Cork, Ireland, at a public meeting "convened for the purpose" in January 1863.
Queens Drive	The pass along the western face of the Remarkables below Double Cone, from south to north. Named by common usage.
Rees River	After and by W. G. Rees, first occupier of land in the Wakatipu, 1860.
Remarkable Mts.	For their unusual character, by Alexander Garvie, surveyor, from Maniototo, 1857.
Shotover	Named by Rees after Shotover House, England, home of G. Gammie, Rees's partner. Called "The Tummel" earlier by R. and A. Cameron.
Skippers	After Malcolm Duncan, miner and former seafarer. Named by common usage.
Slate Range	Phonetic spelling of the Sleat hills, Isle of Skye, Scotland, clan land of the Macdonalds. Possibly named by A. A. Macdonald, 1859.
Tucker Beach	Named because of its lack of gold. (Miners could earn tucker only.) Named by common usage.
Tyndall, Mt.	After a British scientist 1820-93. Named by James McKerrow, 1863.
Von River	After Nicholas von Tunzelmann. Named by James McKerrow, 1863.
Wakatipu, Lake	A Maori name, most probably meaning Trough of the Monster (see Appendix I). Named by early Maoris.

APPENDIX III

PACK TRACKS OF THE WAKATIPU

ALL UNKNOWINGLY, Rees and von Tunzelmann were to follow one of the old Maori trails into the Wakatipu — after them, with the discovery of gold in October-November of 1862, came the flood of miners.

From a population of twelve souls, in less than six months, an estimate made on 18 June 1863, revealed the presence of 8400 miners within the watershed of the Wakatipu basin alone — an astounding increase when one considers the totally undeveloped access routes.

Into the wide tussock slopes, where Rees's infrequent comings and goings had left little or not trace, were cut, almost overnight, the deep trails that are still visible today. Muddy, dangerous and difficult in the extreme, they must have been a nightmare to travel. Almost immediately the Provincial Government was called upon to conduct improvements to the tracks, and such were the pressures of the times that roads were being forced through at the same time as pack tracks were being upgraded to the same destination.

The accompanying sketch shows the major pack tracks with the dates they were superseded by roads — there were some 213 miles of major tracks in the Wakatipu basin.

1 Cardrona Valley: This route was first used by Rees and von Tunzelmann. Connecting with Tobins Track, it led to the Arrow diggings. A road (Crown Range road) was formed through in 1884. An old Maori trail.

2 "Gentle Annie" track from the Kawarau River to the Arrow via the Crown Terrace. It was used until road and punt ferries at Victoria and Morven provided better access in 1864.

3 Kawarau River south side.

4 Macetown-Mototapu. Roses Saddle, 4200 feet.

5 Arrowtown-Macetown (river track). Road started 1881, finished 1884.

6 Arrowtown-Macetown via German and Big Hill, 3400 feet. Specially narrow tracked drays/waggons were used over this track until the river road was opened.

7 Arrowtown-Glendhu (Wanaka) via Mototapu Valley, old Maori trail. Though developed for miners, the track was mostly used by station owners.

8 Arthurs Point-Maori Point, built 1863. Realigned — Long Gully/Deep Creek, 1864. Maori Point-Skippers was a private track with toll bridges at Maori Point and Stony Creek. The east side track was extended to the first Skippers bridge when it was opened in 1866. Skippers road started in 1884, eventually reaching the old bridge in 1888. Skippers Saddle, 3100 feet.

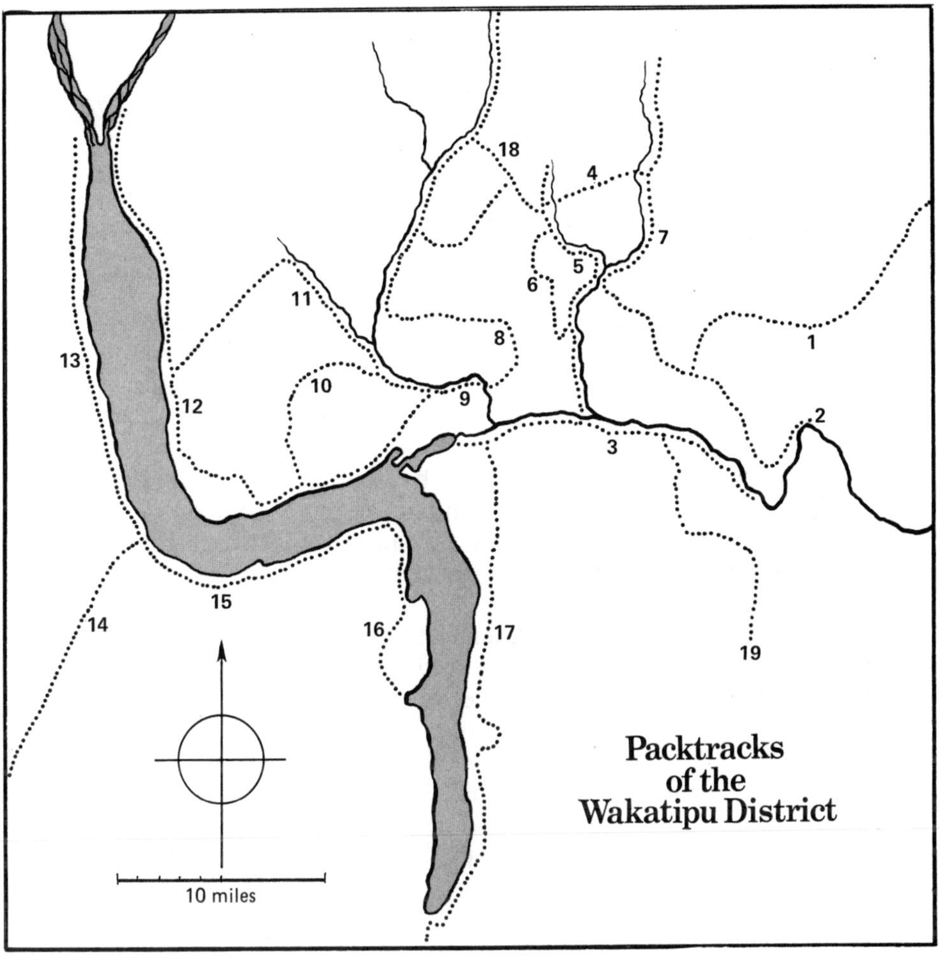

9 Queenstown-Arthurs Point road, 1863.
10 Arthurs Point-Moke Lake.
11 Moonlight-Mount Creighton Station. When track down 25-mile was closed travellers had to go over a 4500-foot ridge.
12 Queenstown-Glenorchy, 1870. Road opened 1962.
13 Kinloch-Greenstone-Mt Nicholas.
14 Mt Nicholas-Mavora. Road first opened in 1929.
15 and 16 Mt Nicholas-Walter Peak-Cecil Peak-Halfway Bay.
17 Kingston-Kawarau River, 1863. Road opened 1936.
18 Macetown-Upper Shotover via Advance Peak saddle, 5500 feet.
19 Gibbston-Nevis, 3500 feet. An old Maori trail.

OTHER REED BOOKS

Sixty Thousand on the Hoof by Peter Newton. A most readable account of station life on sixty stations in the southern half of the South Island, from Mount Peel and Mesopotamia on the Rangitata to Beaumont at the bottom of the island.

Big Country of the South Island by Peter Newton. The shearers and cooks, shpeherds and managers, womenfolk and many more on some hundred stations north of the Rangitata. Also a very useful reference book.

In the Wake of the Axe by Peter Newton. The author turns his attention to the great sheep and cattle stations of the North Island. Fascinating stories recorded with zest and accuracy.

Molesworth by L. W. McCaskill. The story of this great sheep station and its associated properties, traced back to the 1850s. A unique study of the problems of farming New Zealand high country.

Gold Trails of Otago by June A. Wood. A traveller's guide to the goldfields of Otago with an account of the methods used on the goldfields and a compendium of notable and notorious personalities.

QUEENSTOWN
Ben Cruachan
State
Coal Pit Road
Highway No 6
Mt Edward
Mt Rosa
THE REMARKABLES
LAKE
WAKATIPU
Lake Alta
Double Cone
Mt Salmond
Doolans Creek
Branch)
Doolans Creek (Left Branch)
Wye Creek
Nevis Burn
South Br
Ben Nevis
Schoolhouse Creek
James Peak
Garston Nevis Road
THE REMARKABLES
AND
MOUNTAINS
Date of Photography Feb 73
CAUTION
This mosaic is not an accurate map
0 2 4 6 10 Km
Survey Mosaic No 591

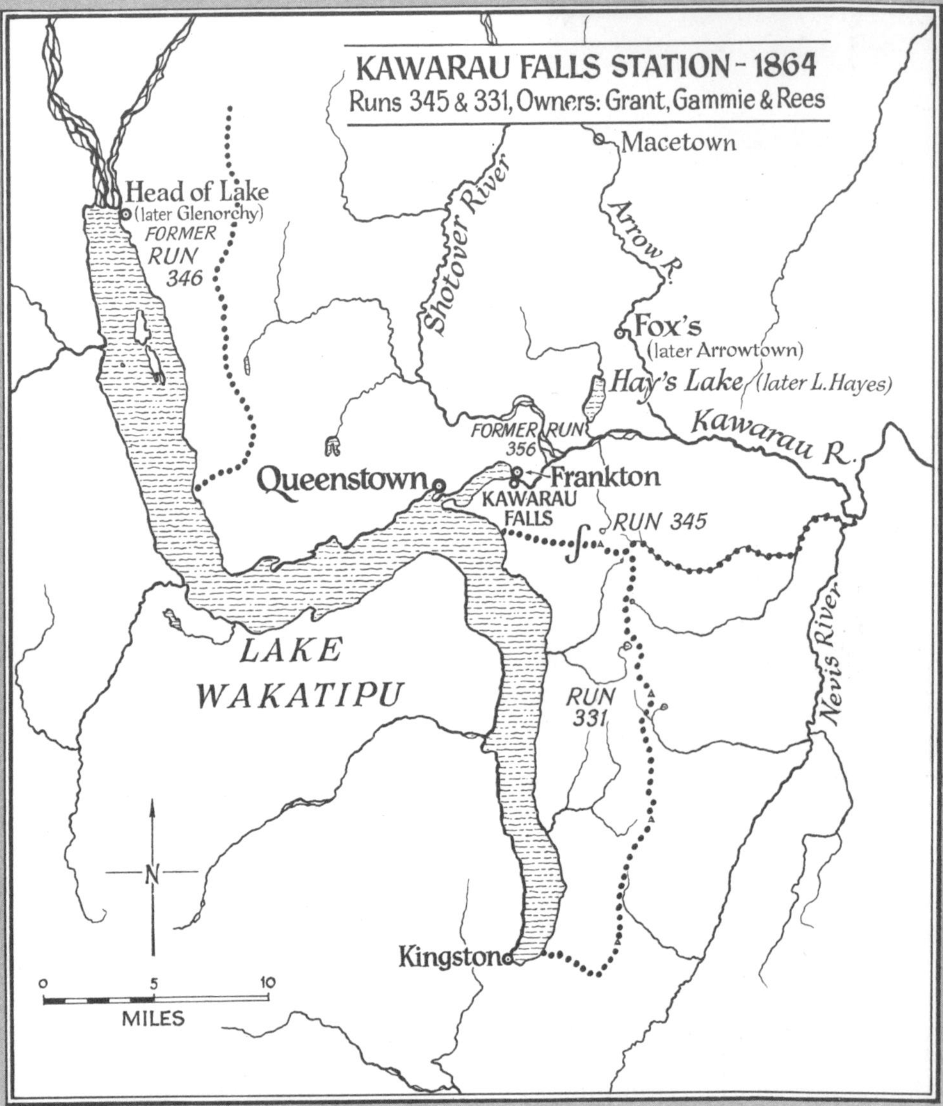
KAWARAU FALLS STATION - 1864
Runs 345 & 331, Owners: Grant, Gammie & Rees
Macetown
Head of Lake
(later Glenorchy)
FORMER
RUN
346
Shotover River
Arrow R.
Fox's
(later Arrowtown)
Hay's Lake (later L. Hayes)
Kawarau R.
FORMER RUN
356
Queenstown
Frankton
KAWARAU
FALLS
RUN 345
LAKE
WAKATIPU
RUN
331
Nevis River
N
Kingston
0
5
10
MILES